T0133397

Bibliografische Information der Deutschen Nationalbibliothek

Die Deutsche Nationalbibliothek verzeichnet diese Publikation in der Deutschen Nationalbibliografie; detaillierte bibliografische Daten sind im Internet über http://dnb.d-nb.de abrufbar.

ISBN 978-3-8325-2397-8

Logos Verlag Berlin GmbH
Comeniushof, Gubener Str. 47,
10243 Berlin
Tel.: +49 (0)30 42 85 10 90
Fax: +49 (0)30 42 85 10 92
INTERNET: http://www.logos-verlag.de

Dissertation zur Erlangung des akademischen Grades
doctor rerum naturalium
der Mathematisch–Naturwissenschaftlichen Fakultät
der Universität Rostock

Thermodynamics of Low-dimensional Light Front Gauge Theories

vorgelegt von
Dipl. Phys. Stefan Strauß

January 20, 2010

AG Elementarteilchenphysik
Institut für Physik · Universität Rostock

Dissertation at Rostock University

Referees

1. PD Dr. Michael Beyer, University of Rostock, Rostock, Germany
2. Prof. Stanley J. Brodsky, SLAC National Accelerator Laboratory, Stanford University, Menlo Park, USA
3. Prof. Tobias Frederico, Centro Técnico Aeroespacial, São José dos Campos, Brazil

Date of Oral Presentation

October 12, 2009

Contents

Abstract

In this thesis a consistent non-perturbative application of light cone quantization for the thermodynamics of strongly coupled quantum field theories is presented. Non-perturbative methods are mandatory for a description of physical systems under extreme conditions such as the fire ball occurring during heavy ion collisions or the plasma of the early universe.

Using the general light cone frame, it is shown that thermodynamic properties can be meaningfully determined in the framework of light cone quantized Hamiltonian field theory. This work focuses on Quantum Electrodynamics (QED_{1+1}/massive Schwinger model) and Quantum Chromodynamics in $1+1$ dimensions to compute thermodynamic observables in an *ab initio* approach. The central quantity is the partition function $\mathcal{Z}$ used to derive all other observables, e.g. the equation of state, by standard relations.

Due to the increased numerical effort and new solution algorithms the accuracy on low lying bound state masses of the massive Schwinger model could be improved by almost two orders of magnitude compared to previous light cone calculations. Finally, a possible application of the density matrix renormalization group to the massive Schwinger model has been explored.

Zusammenfassung

Gegenstand der vorliegenden Arbeit ist die Entwicklung einer konsistenten Beschreibung stark wechselwirkender relativistischer Quantenfeldtheorien bei endlichen Temperaturen. Dies wurde durch konsequente Anwendung der Lichtkegelquantisierung realisiert, die insbesondere geeignet ist nichtstörungstheoretisch beschreibbare Phänomene zu behandeln.

Nichtstörungstheoretische Untersuchungen sind notwendig für physikalische Systeme unter extremen Bedingungen, wie sie im frühen Universum oder in dem Feuerball einer Schwerionen-Kollision vorherrschen. In der Arbeit wurde gezeigt, daß es möglich ist, mit Hilfe der verallgemeinerten Lichtkegelquantisierung sinnvolle thermodynamische Observablen zu bestimmen.

Ausgehend von einer grundlegenden Theorie – es wurde die Quantenelektrodynamik (QED_{1+1}/massives Schwingermodell) und die Quantenchromodynamik in 1+1 Dimensionen behandelt – sind *ab-initio*-Berechnungen thermodynamischer Größen durchgeführt worden. Die zentrale thermodynamische Größe ist dabei die Zustandssumme $\mathcal{Z}$, aus der weitere Größen, z.B. die Zustandsgleichungen, abgeleitet wurden.

Durch neu entwickelte Lösungsalgorithmen konnte in dem Zusammenhang auch die Genauigkeit der Massen und Strukturfunktionen der isolierten Bindungszustände in der QED_{1+1} im Vergleich zu vorherigen Rechnungen deutlich verbessert werden. Weiterhin ist eine mögliche Anwendung der Dichtematrix-Renormierungsgruppe im Schwingermodell auf dem Lichtkegel untersucht worden.

1. Introduction

The fundamental forces of nature are the following four: electro-magnetic, weak and strong interaction as well as gravity. The first three are described by gauge theories of simply connected, compact, non-abelian Lie groups. General relativity can be regarded as a gauge theory of the diffeomorphism group. Although the quantum formulation of these fundamental theories is worked out (except for gravity), one is hardly able to compute observables without using perturbation theory. However, many physical effects are suspected beyond perturbative treatment in strongly coupled theories and non-perturbative methods have to be developed. The prototype example of such a strongly coupled theory in particle physics is Quantum Chromodynamics (QCD), based on the gauge group SU(3), at low energies. Its most important properties in this energy regime are color confinement and chiral symmetry breaking. Due to confinement quarks do not appear as physical particles in the theory. Only hadrons, color-singlet bound states of quarks, are observable. Chiral symmetry breaking is usually connected to the dynamical mass generation by QCD condensates, responsible for the majority of the mass we experience in the everyday world. In common notion, all these properties are thought to be caused by the complicated QCD ground state that nobody has determined so far. At high energies or high momentum transfers quarks and gluons can be regarded as nearly free because the running QCD coupling becomes a small parameter (asymptotic freedom). This has allowed to demonstrate the existence of quarks in high energy scattering experiments. The possibility to observe free quarks and gluons is also given if a hadron gas is heated up to temperatures higher than the critical temperature $T_c \sim 170$ MeV at vanishing quark chemical potential, the quark gluon plasma (QGP). The transition from hadronic matter to quark matter is called the confinement-deconfinement transition. Besides the QGP and the hadronic phase there are other phases expected to be present at low temperatures and large chemical potentials, like the color superconducting (CSC) and color flavor locked (CFL) phase. A schematic view of the whole quark matter phase diagram is shown in Figure 1.1.

The investigation of these extreme matter phases are the subject of various experiments. Among these are heavy ion collisions of Au nuclei presently performed at RHIC[1]. The RHIC collider operates at a center of mass energy of $\sqrt{s} = 250$ GeV/nucleon . The first results from collaborations working at RHIC indicate the existence of the QGP as an almost perfect fluid, opposed to the expectations of a ideal gas of quark and gluons due to asymptotic freedom [2]. Previously, hints on the QGP were also found at the SPS[2] that collided Pb nuclei at an energy of 160 GeV/nucleon. The expansion and dilution of the QGP can be successfully described by

[1] Relativistic Heavy Ion Collider at the Brookhaven National Laboratory

[2] Super Proton Synchrotron at CERN

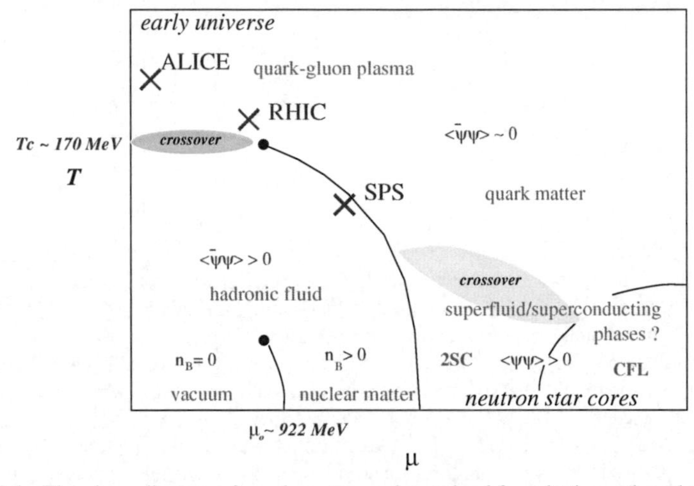

Figure 1.1.: The phase diagram of quark matter as determined from lattice and model computations, taken from [1]. The black lines are phase transitions of first order while the crossovers are indicated as blue and yellow patches. The black dot is the critical endpoint. Blue crosses depict the different regions of the phase diagram where the present and future experiments (will) operate.

(viscous) relativistic hydrodynamics, see e.g. [3]. This is rather surprising since the thermal system created in a nucleon-nucleon collision only consists of several thousand particles and is hardly a macroscopic system. However, a hydrodynamical description is sensible because experiments have shown the formation of collective phenomena, like elliptic flow, and the mean free path within the QGP is small compared to the total size of the system. Important input for the hydrodynamical computations are the initial conditions shortly after the collision and the thermodynamical equation of state of the strongly coupled system.

Several heavy ion collision experiments are planned for the future. Namely, the Alice[3] experiment at the LHC[4] will be run in 2012 and the CBM[5] experiment at FAIR[6] in 2013. While at the LHC will collide Pb nuclei at the huge energy of $\sqrt{s} = 2.5$ TeV/nucleon, the FAIR ring aims at lower center of mass energies of 34 GeV/nucleon. Therefore Alice will be able to explore the high temperature crossover, opposed to CBM which is supposed to investigate the low temperature, large density regime. Yet other physical systems, where the equation of state for quark matter is important, are compact stellar objects in astro-particle

[3]A Large Ion Collider Experiment
[4]Large Hadron Collider at CERN
[5]Compressed Baryonic Matter
[6]Facility for Antiproton and Ion Research at GSI Darmstadt

physics and the early universe. One conjectures the presence of a quark core in the highly dense interior of compact neutron stars. Compact stars are rather cold and dense systems, thus the relevant quark matter phase are the CSC and CFL phase. This can be observed indirectly in the properties of the star like the cooling behavior and thermal emissions. It is understood that for these physical applications the determination of the whole quark matter phase diagram from first principles is highly desirable.

Non-perturbative field theory is mostly studied in the lattice regularization of the path integral representation accompanied by the Monte-Carlo (MC) sampling method and the real-space renormalization group. For QCD one was able to compute various quantities like the quark-antiquark potential showing confinement at large coupling, particle spectra, thermodynamical quantities, form factors using the lattice gauge theory. Besides the overwhelming success of lattice gauge theory, it also has some shortcomings. Most notably, there are the theoretical cases where the MC algorithm fails numerically because of the sign/complex action problem. Famous instances of this problem are the inability of lattice gauge theory to explore the $\mu \geq T$ part of the QCD phase diagram or if non-trivial topological terms are present in the action. The generic Monte-Carlo sign problem is at least as hard to compute as problems in the complexity class NP (class of non-deterministic polynomial problems) and every problem in NP is reducible to the sign problem in polynomial time (i.e. in short words the generic Monte-Carlo sign problem is NP-hard) [4]. Additionally, some quantities are rather cumbersomely extracted from lattice computations, e.g. mass spectra as exponential scaling of correlators in Euclidean times or any physics involving time-like processes. Besides, all quantities are computed in Euclidean space-time and should be analytically continued to Minkowski space when necessary. Although the mass gap in pure Yang-Mills theory has been successfully calculated in the context of the action-based lattice approach, a Hamiltonian-based analysis could address such questions more directly and also deliver the wavefunctions necessary for a complete description of the glueballs.

The present thesis utilizes the light front (LF) Hamiltonian approach to quantum field theory. Light front field theory bases on Dirac's front form of relativistic Hamiltonian dynamics introduced in Ref. [5]. The theory is quantized at a hyper-plane tangent to the light cone, i.e. at equal light cone times. Many features of this approach are distinct from the standard instant form where the theory is quantized on space-like hyper-planes. Among these are the constrained quantization that reduces the independent degrees of freedom, and a Fock space expansion build on the different, simpler ground state structure. The simplicity of the light front vacuum state can be justified by the fact that one of the kinematical light cone momenta is always non-negative. A further advantage is the large number of kinematical Poincaré generators in the front form, including the boost generator in z-direction, that allow the computation of frame-independent and process-independent wavefunctions of bound states. Knowing the LF wavefunctions one is able to directly make contact to observables, like the parton structure of hadrons, that are testable in experiments. Light front quantized field theory was advocated, as a complement to Monte Carlo simulations in lattice gauge theory, among others [6] by the Nobel laureate Ken Wilson [7];

I also worry that there is not enough research on approaches to solving QCD that could be complementary to Monte Carlo simulations, such as the lack of any comparable research build-up on light-front QCD... My third concern is to suggest that there needs to be more attention to research on light-front QCD as a complement to research on lattice gauge theory.

A genuine non-perturbative approach to light cone field theory is discrete light cone quantization (DLCQ). A momentum space lattice is introduced by compactifying a light-like direction. By considering Fock space sectors of increasing total momentum one derives a matrix representation of the light cone Hamiltonian H_{LC}. The theory can be solved by considering a Schrödinger-type matrix equation

$$H_{\mathrm{LC}}|\Psi\rangle = M^2|\Psi\rangle. \tag{1.1}$$

Generalizing light front field theory to finite temperatures is not straightforward. Usually, the physical theory is quantized in the rest system of the medium and the medium four-velocity coincides with the four-vector assigning the time direction. Temperature is introduced into the theory by compactifying the time direction to a circle of length $\beta = 1/T$. This does not work with a light-like evolution direction because the medium can not be boosted to the light front frame. The general light front frame [8] changes the transformation to light front slightly and mixes light front and instant form coordinates. In a special case the time evolution is driven by the LF energy operator $P^0_{\mathrm{LC}} = \frac{1}{2}(P^+ + P^-)$. Note that the LF energy operator is in general different from the instant form Hamiltonian since P^0_{LC} is given in terms of LF initialized or quantized fields.

Although our main motivation is the phase diagram of quark matter we consider here the thermodynamics of light front gauge theories in 1+1 dimensions. The reason is simply that the computational task of non-perturbative light front QCD in 3+1 dimensions is too demanding. On the contrary for low-dimensional systems DLCQ computations for diverse theories have proven to provide reliable results spending modest computational resources. Here we focus on the spectrum and the thermodynamics of the massive Schwinger model or Quantum Electrodynamics in 1+1 dimensions and QCD_{1+1}. In particular, the Schwinger model has many properties familiar from QCD in four dimensions. These are e.g. charge confinement, topological excitations, dynamical mass generation by chiral symmetry breaking, and the axial anomaly. Investigating these simplified models in light front quantization helps clarifying which physical effects are depending on the quantization scheme. When the LF Hamiltonian is at hand the thermodynamics of these models is found by computing the partition function

$$\mathcal{Z} = \mathrm{Tr}\, e^{-\frac{\beta}{2}(P^+ + P^-)}. \tag{1.2}$$

Further quantities can be derived directly from the partition function using thermodynamical relations.

This work is organized as follows.

In Section 2 the notion of light front quantization will be explained in more detail. In particular, quantization in a finite box is introduced and problems connected to zero modes are mentioned.

The generalization of light front field theory to finite temperatures is treated in Section 3. The light front statistical operator is related to the standard relativistic statistical physics and thermodynamics. We present the foundation of (perturbative) thermal field theory in the general light front frame. The thermodynamics of an ideal fermionic gas is examined and the numerical results are compared to the textbook results.

These methods are applied in Section 4 to the massive chiral light front Schwinger model at zero and finite temperature. Several properties of the Schwinger model are summarized and their implementation on the light front is discussed. Numerical results for mass spectra, valence wavefunction, and structure functions for various couplings are given with improved accuracy. The main results are the thermodynamical quantities, such as the pressure, internal energy, entropy and the equation of state, computed for the LC Schwinger model in Section 4.4.

In Section 5 the same techniques are used to determine the thermodynamical potential for QCD_{1+1}. The crucial differences to the LF Schwinger model computations are outlined.

Finally, the sixth Section outlines the application of the renormalization group to discrete light cone quantization. After reviewing briefly the traditional similarity renormalization group approach, we focus on relating the density matrix renormalization group of spin chain systems to one-dimensional light front field theory.

In the last Section we summarize our findings and point out future perspectives. Many technical details are given in Appendices and referred to in the main text, when needed.

Throughout the thesis we work in high energy physics units, i.e. all quantities are given in units of $\hbar, c, k_B$. Note that the terms 'light cone' (LC) and 'light front' (LF) are used interchangeably. We collected the most important definitions and notations in Appendix A.

2. Light front quantization

In the first chapter we shall explain the notation of light front quantization. This will be done in the context of Dirac's characterization of different forms of relativistic dynamics [5]. The subsequent sections are then dealing with discrete light cone quantization (DLCQ) which has initially been developed to treat the zero mode problem [9]. DLCQ is a truly non-perturbative approach to quantum field theory. It also allows us to compute the mass spectra and light front wavefunctions using the LF Fock space expansion. Quantization can be treated by the Dirac-Bergman formalism of constrained Hamiltonian systems as well as functional approaches like the Faddeev-Jackiw construction. We will outline the latter in the case of scalar fields. This section mainly refers to the common introduction and review articles on LF quantization [10, 11, 12, 13].

2.1. Light front form and other forms of relativistic dynamics

There are many different causal foliations of Minkowski space-time and each of them gives rise to a formulation of relativistic theories. By choosing a certain foliation, one specifies a time evolution parameter and a hyper-surface Σ where the initial conditions of the equations of motion are defined. The generic hyper-surface Σ is further restricted to contain no time-like direction locally [13], i.e. the normal vector $N^\mu(x)$ of Σ is time- or light-like. Demanding the stability group of the surface Σ, that is the subgroup of Poincaré transformations leaving Σ invariant, to be transitive (i.e. Σ is a single orbit under the action of the stability group) reduces the number of possible forms of dynamics drastically [14]. The outcome analyzing all subgroups of the Poincaré group are the five relativistic forms of dynamics listed below

- Instant form $\Sigma : x^0 = 0\,, d = 6,$
- Light front form $\Sigma : x^0 + x^3 = 0\,, d = 7,$
- Point form $\Sigma : {x^0}^2 - \boldsymbol{x}^2 = a^2 > 0, \quad x^0 > 0\,, d = 6,$
- First hyperboloid form $\Sigma : {x^0}^2 - x_\perp^2 = a^2 > 0, \quad x^0 > 0\,, d = 4\,,$
- Second hyperboloid form $\Sigma : {x^0}^2 - x_1^2 = a^2 > 0, \quad x^0 > 0\,, d = 4.$

Here d denotes the number of generators of the stability group of Σ and a is an arbitrary, nonzero constant. Note that the front form has the maximum number of generators. The last

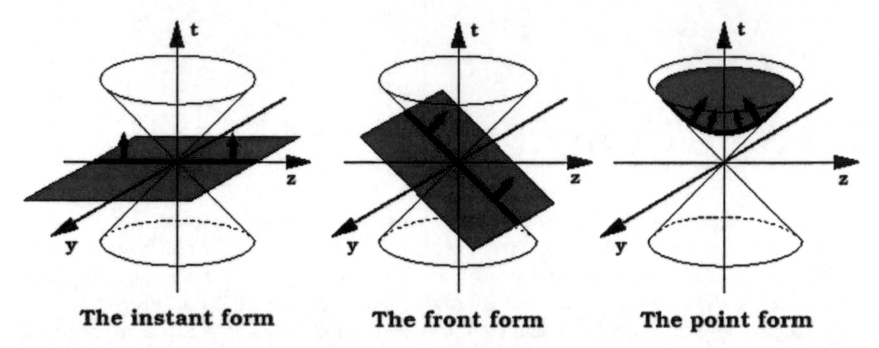

Figure 2.1.: Shown are of the most commonly used forms of relativistic dynamics taken from Ref. [10]. On the hyper-surfaces that are colored in red the initial conditions in the classical theory or the equal time commutators in the quantum theory are given.

two forms have a very reduced stability group and are usually not considered. We will later in this section specify the elements of the stability group of the light front form more closely.

The three most common choices are depicted in Figure 2.1. The standard choice of the time direction and Hamiltonian is the instant form (left side of Figure 2.1). Only very limited work has been devoted to constructed theories in the point form (right side of Fig. 2.1), e.g. [15]. This work is devoted to the light front form (middle of Figure 2.1). Thus all initial conditions in the classical theory and commutation relation in the quantum theory are chosen at the equal light front surface $\Sigma : x^+ = x^0 + x^3 = 0$. Following the notation of Ref. [10], the light front coordinates can be introduced

$$\begin{aligned} x^+ &= x^0 + x^3, \\ x^- &= x^0 - x^3, \\ x_\perp^i &= x^i \,, \qquad i = 1, 2. \end{aligned} \tag{2.1}$$

The $x^+(x^-)$-component is referred to as light front time (space). Under the transformation to LC coordinates (2.1) the metrics alter to a off-diagonal form;

$$\eta_{\mu\nu} = \begin{pmatrix} 0 & 1/2 & 0 & 0 \\ 1/2 & 0 & 0 & 0 \\ 0 & 0 & -1 & 0 \\ 0 & 0 & 0 & -1 \end{pmatrix}, \qquad \eta^{\mu\nu} = \begin{pmatrix} 0 & 2 & 0 & 0 \\ 2 & 0 & 0 & 0 \\ 0 & 0 & -1 & 0 \\ 0 & 0 & 0 & -1 \end{pmatrix}. \tag{2.2}$$

Obviously the transformation to light front coordinates (2.1) is not a Poincaré transformation. The conjugate variables to (2.1) fulfill the following dispersion relation

$$k^2 = k^+k^- - k_\perp^2 = m^2 \quad \Rightarrow \quad k^- = \frac{m^2 + k_\perp^2}{k^+} \quad \text{for} \quad k^+ \neq 0. \tag{2.3}$$

Conventionally, the momentum component k^- is called light cone energy while k^+ is the longitudinal momentum (sometimes referred to as LF mass) and $k_\perp$ are the transverse kinematical momenta. Note that both k^+ and k^- are always non-negative for physical particles, i.e. obeying the conditions $k^2 \geq 0, k^0 \geq 0$.

There are several physical systems or high-energy processes where light front quantization is particularly useful. Most prominently are deep inelastic hadron-lepton scattering experiments (DIS). These experiments probe the structure of the hadron and measure the structure functions revealing the constituents of hadrons. When one of the constituents is struck it receives a high-momentum transfer and travels nearly with the speed of light along the light cone. In the Bjorken limit the dominating contribution comes from the case that only one quark in the hadron is coupled to the exchange photon (the so-called handbag diagram). The struck quark does not interact with the other constituents (in light cone gauge) and one can imagine it being removed at one place and reappearing at a light-like distance away. Therefore the DIS experiments determine the light-like correlation functions

$$\langle\psi(P)|\bar{\psi}(x^-, x_\perp)\psi(y^-, y_\perp)|\psi(P)\rangle \tag{2.4}$$

of the target, where $|\psi(P)\rangle$ denotes the state vector of the hadron having momentum P. In light front quantization the frame independent state vector can be computed solving the light cone time independent Schrödinger equation

$$H_{\mathrm{LC}}|\psi(P)\rangle = M^2|\psi(P)\rangle, \tag{2.5}$$

with the mass eigenvalue M. Only the light cone framework is able to compute the light front correlation functions (2.4) as static, ground state properties of the relativistic bound state. Other techniques would require to solve dynamical equations, i.e. the time evolution of the state. In this sense the LC wavefunction is the natural quantity to describe the hadron state in the context of high-energy scattering. The multi-parton Fock space expansion of the state is

$$\begin{aligned} |\psi(P)\rangle = \sum_n \int \left[\prod_{i=1}^{n} d^2k_{\perp i} dx_i\right] \delta(P_\perp - \sum_i k_{\perp i})\delta(1 - \sum_i x_i) \\ \sum_{s_i} \psi_n(x_i, k_{\perp i}, s_i)|n; x_1, k_{\perp 1}, \ldots, x_n, k_{\perp n}, s_i\rangle \end{aligned} \tag{2.6}$$

where $x_i, k_{\perp i}, s_i, n$ stand for the momentum fraction, transverse momentum, helicity of the i-th parton and the n particle sector, accordingly. Additionally, ψ_n is the LC wavefunction in the n particle sector. This Fock space expansion in free quark states is expected to converge fast. Hence, it describes the dominance of valence quarks in the hadron and makes contact with the phenomenological successful constituent quark models. Sophisticated renormalization methods have been developed, see Section 6, with the aim to obtain accurate results by expanding (2.6) only to few-body sectors. The parton distribution function is directly given by the norm squared of the amplitudes ψ_n on the rhs of (2.6) and is connected to the famous Bjorken scaling functions F_1, F_2 in DIS.

Light front quantization is in close connection to the so-called infinite momentum frame (IMF), see [16]. The IMF formalism was used in the so-called old-fashioned Hamiltonian perturbation theory where it was realized that certain diagrams vanish when the external four-momentum is taken to infinity [17]. Loop diagrams where the momentum of the external lines is taken to be zero (vacuum diagrams) give vanishing contribution. This fact supports the idea of a simplified vacuum structure in light front quantization. The vacuum is trivial in the sense that the ground state of the interacting theory is the vacuum state of the free theory, thus contains no quanta. In a generic quantum theory the vacuum state $|0\rangle$ solves

$$\hat{P}^{\mu}|0\rangle = 0. \tag{2.7}$$

Since $\hat{P}^{+}$ is kinematic, the operator contains no interaction and is simply given by the sum of single particle p^{+} momenta. Any particle excitation $a^{\dagger}(k^{+}, k_{\perp})|0\rangle$ contributes positive momentum k^{+}. The total momentum P^{+} is conserved in the interacting theory and the unique state having the quantum numbers (2.7) (and maybe additionally other ones) is the kinematical Fock vacuum, having vanishing particle number. Moreover, this only holds if one excludes zero modes considered in Section 2.4. Nevertheless, the triviality of the ground state in LC quantized field theories presents a great conceptional and technical simplification that makes LF quantization a promising and successful approach to non-perturbative physics in strongly-coupled field theories.

The question arises how interesting effects like confinement, chiral symmetry breaking of QCD usually connected to complicated vacuum structure in the instant form are realized in light cone quantization. Zero modes of the gauge field are regarded to be responsible for these phenomena. Thus, treating zero mode degrees of freedom correctly leads to a non-trivial vacuum, or more complicated, symmetry phase dependent operators. Many attempts have been pursued to solve this question, some ideas will be reviewed in later sections. We conclude this overview section by mentioning three appealing features of the light front form in some detail:

1. Front form physical systems exhibit similarities to non-relativistic physical systems [13, 12] which in fact makes it possible to separate the Hamiltonian into parts capturing the c.m. movement and internal ones. Such a separation is generally not possible in the instant form. This is easily seen when one examines the Poincaré algebra

$$\begin{aligned} \{P^{\mu}, P^{\nu}\} &= 0, \\ \{M^{\mu\nu}, P^{\sigma}\} &= g^{\nu\sigma}P^{\mu} - g^{\mu\sigma}P^{\nu}, \\ \{M^{\mu\nu}, M^{\varrho\sigma}\} &= g^{\mu\sigma}M^{\nu\varrho} - g^{\mu\varrho}M^{\nu\sigma} - g^{\nu\sigma}M^{\mu\varrho} + g^{\nu\varrho}M^{\mu\sigma}. \end{aligned} \tag{2.8}$$

Here P^{μ} are the generators of space-time translations while $M^{\mu\nu}$ denotes angular momentum and boost generators. The brackets $\{\cdot,\cdot\}$ in (2.8) are Lie brackets. See Appendix A for the explicit representation of the Poincaré generators for field theories.

We like to focus one a certain subalgebra of (2.8)

$$\begin{aligned} \{M^{12}, M^{+i}\} &= \varepsilon^{ij} M^{+j}, \\ \{M^{12}, P^{i}\} &= \varepsilon^{ij} P^{j}, \\ \{M^{+i}, P^{-}\} &= -2P^{i}, \\ \{M^{+i}, P^{j}\} &= \delta^{ij} P^{+}, \end{aligned} \tag{2.9}$$

where ε^{ij} is the anti-symmetric symbol in two dimensions and all missing relations to close (2.9) have trivial rhs. The algebra (2.9) is isomorph to the algebra of the Galilei group in two dimensions due to the following identifications: $P^- \leftrightarrow H$, $P^i \leftrightarrow k^i$, $P^+ \leftrightarrow 2m$, $M^{i+} \leftrightarrow -2G^i$, $M^{12} \leftrightarrow L$. The Galilei group is generated by the Hamiltonian H, space translations k^i, Galilean boosts $G^i = mx^i$, angular momentum L and m is the Casimir operator of (2.9). Because of the mapping of the subgroup to the Galilei group one expects, to some extent, non-relativistic features in the light cone frame. An example is the form of the LC Hamiltonian

$$H = \frac{k_i^2}{2m} + E_{\text{int}} \leftrightarrow \frac{P_i^2}{2P^+} + \tilde{E}_{\text{int}}, \tag{2.10}$$

which is the sum of free part and the Galilei invariant interaction E_{int}. See also the discussion on light-like compactifications in [18] .

2. Another advantage of LC quantization is the maximal number of kinematical Poincaré generators: In, e.g., a matrix representation on Minkowski space of (2.8) the generators leaving the surface $\Sigma : x^+ = 0$ invariant are: $\{P^+, P^i, M^{+1}, M^{+2}, M^{+-}, M^{12}\}$. Since $\Sigma : x^+ = 0$ is a 3-dimensional hyper-plane one naturally expects six kinematical generators, namely three translation in the plane and three rotations. Yet an extra generator compared to instant and point form is available because $\Sigma : x^+ = 0$ is light-like and as such boost transformation generated by M^{+-} in x^3-direction are preserving Σ. The x^3-boost is in x^+, x^- simply a scale transformation and a kinematical operation in $x_\perp$;

$$\begin{aligned} x^+ &\longrightarrow e^{\omega} x^+, \\ x^- &\longrightarrow e^{-\omega} x^-, \\ x_\perp^i &\longrightarrow x_\perp^i + \beta^i x^+ \end{aligned} \tag{2.11}$$

with rapidity ω and arbitrary parameters β^i. By using (2.11) in momentum space one is able to compute frame-independent wavefunctions for bound states [10]. See also Section 4.3.

3. The last point we would like to mention is a method called near light cone (NLC) quantization. It was first proposed in Ref. [19]. One introduces slightly modified LC coordi

nates

$$
\begin{aligned}
x^+_\varepsilon &= x^+ + \frac{\varepsilon}{L} x^-, \\
x^-_\varepsilon &= x^-, \\
x^i_\perp &= x^i,
\end{aligned}
\tag{2.12}
$$

where x^+, x^- are given by (2.1). Quantization in NLC coordinates is done on a space-like plane $\Sigma_\varepsilon : x^+_\varepsilon = 0$ and therefore for $\varepsilon \neq 0$ equivalent to a boosted instant form approach. Many problems peculiar to light front form, like possible violation of micro-causality [13, 20], constrained quantization and zero modes (see next sections), are avoided within this approach. Nevertheless, it is not clear whether in the limit $\varepsilon \longrightarrow 0$ the NLC theory goes smoothly for every observable to the light front counterpart (some observables where the $\varepsilon \longrightarrow 0$ limit in well-behaved are given in [19]). There has been work in perturbative scalar field theory [21] indicating the opposite. For a recent application of NLC QCD in 3+1 dimensions, see [22].

2.2. Quantizing light front field theories

The term 'Light front quantization' usually denotes canonical quantization of Hamiltonian field theory based on initial conditions of the field variables on the light front plane $\Sigma : x^+ = 0$. Canonical quantization means that the Poisson structure of functions on phase space (observables) of the classical theory is identified with the commutator algebra of (hermitian) operators in quantum theory.

For canonical quantization one demands commutator relations only for the independent degrees of freedom, thus any constraints have to be dealt with before. But also within the classical theory it is advantageous to adapt to natural variables which fulfill the constraints automatically. Light front quantization has two issues here which can not clearly be separated from each other. First, systems described in LF coordinates have singular Lagrangians that are first-order in the x^+ derivative and as such their canonical momenta are constrained. And secondly, handling the various zero modes imply additional constraints. In this section we are concerned with the first issue and postpone the second to Section 2.4.

Scalar fields

Let us consider for simplicity massive scalar fields in four dimensions

$$
\mathcal{L} = \frac{1}{2}\partial_\mu\phi\,\partial^\mu\phi - \frac{m^2}{2}\phi^2 - V(\phi), \tag{2.13}
$$

where the prototype example would be to take $V(\phi) = \frac{\lambda}{4}\phi^4$. Rewritten in LC coordinates one recognizes that (2.13) is a first-order system in light cone time

$$
\mathcal{L} = 2\partial_+\phi\partial_-\phi - \frac{1}{2}(\partial_\perp\phi)^2 - \frac{m^2}{2}\phi^2 - V(\phi). \tag{2.14}
$$

The canonical momentum

$$\pi(x) = \frac{\partial \mathcal{L}}{\partial(\partial_+\phi)} = 2\partial_-\phi = 2\frac{\partial\phi}{\partial x^-} \tag{2.15}$$

is *not* an independent variable since it is determined non-dynamically by $\phi(x)$ and Eq. (2.14) should be regarded as a constraint. Following the standard procedure the naive LC Hamiltonian density is

$$\mathcal{H} = \pi\partial_+\phi - \mathcal{L} = \frac{1}{2}\left(\partial^\perp\phi\right)^2 + \frac{m^2}{2}\phi^2 + V(\phi) \tag{2.16}$$

and leads to the Hamiltonian equation of motion

$$\partial_+\pi = 2\partial_+\partial_-\phi = 2\left\{\partial_-\phi, H\right\} = \partial_\perp^2\phi - m^2\phi - V'(\phi), \tag{2.17}$$

where we have used $H = \int_{\mathrm{LC}} d^3x\, \mathcal{H}$ and the Poisson bracket definition for functionals

$$\{F,G\}_{x^+=y^+} = \int_{\mathrm{LC}} d^3x \left(\frac{\delta F}{\delta\phi}\frac{\delta G}{\delta\pi} - \frac{\delta F}{\delta\pi}\frac{\delta G}{\delta\phi}\right). \tag{2.18}$$

Equation (2.17) differs from the Klein-Gordon equation

$$4\partial_+\partial_-\phi - \partial_\perp^2\phi + m^2\phi + V'(\phi) = 0, \tag{2.19}$$

that one would expect from Euler-Lagrange equations of (2.13), by a factor of 2. This conflict between Lagrangian and Hamiltonian equations of motion indicates that we used a not well-suited Hamiltonian to drive the time-evolution. Especially the constraints (2.15) have not been accounted for. There are two ways in the literature to derive a appropriate Hamiltonian or Poisson brackets for constrained physical system: the Dirac-Bergman (see, e.g. the appendix of [10]) or the Faddeev-Jackiw [23] algorithm. Both can of course be carried through for the free LC field theories and we are going to discuss shortly the latter since it is computationally less extensive [24, 25, 26, 13].

The following procedure is well-defined for dynamical systems with a finite number of degrees of freedom and we carry this approach over to field theories here without worrying about technicalities. The aim is to compute the symplectic structure ω of the phase space because the inverse of ω determines the elementary Poisson brackets by

$$\left\{\xi^i(x), \xi^j(y)\right\} = \left(\omega^{-1}\right)^{ij}(x,y), \tag{2.20}$$

where ξ collectively denotes the phase space coordinates $\xi = (\phi_r(x), \pi_r(x))$. One starts by identifying the components of the canonical one-form $\theta_i(\xi, x)$, since the components of the natural symplectic two-form on phase space are defined as

$$\omega_{ij}(x,y) = \frac{\delta}{\delta\xi^i(x)}\theta_j(\xi,y) - \frac{\delta}{\delta\xi^j(y)}\theta_i(\xi,x). \tag{2.21}$$

In the front form the phase space coordinates are reduced to $\xi^i(x) = (\phi(x), 0)$ because the canonical momentum (2.15) is constrained. However, the canonical one-form reads

$$\theta(x) = 2\partial_-\phi(x). \tag{2.22}$$

Inserting (2.22) into (2.21) one finds the LC symplectic two-form

$$\begin{aligned}\omega(x^-, x_\perp, y^-, y_\perp) = \frac{\delta\theta(y)}{\delta\phi(x)} - \frac{\delta\theta(x)}{\delta\phi(y)} &= \left(\partial_{y_-} - \partial_{x_-}\right)\delta(y^- - x^-)\delta^2(x_\perp - y_\perp) \\ &= -2\partial_{x_-}\delta(x^- - y^-)\delta^2(x_\perp - y_\perp).\end{aligned} \tag{2.23}$$

The inverse of ω defined by $\int d^3x'\omega^{-1}(x,x')\omega(x',y) = \delta(x,y)$. Rewriting the delta distributions in (2.23) via exponentials

$$\omega(x^-, x_\perp, y^-, y_\perp) = 2i\int \frac{d^3p}{(2\pi)^3} p_- e^{-ip_-(x^- - y^-) - ip_\perp\cdot(x_\perp - y_\perp)} \tag{2.24}$$

one can read off the inverse

$$\begin{aligned}\omega^{-1}(x^-, x_\perp y^-, y_\perp) &= \frac{-i}{2}\int \frac{d^3p}{(2\pi)^3}\frac{1}{p_-} e^{-ip_-(x^- - y^-) - ip_\perp\cdot(x_\perp - y_\perp)} \\ &= -\frac{1}{4}\varepsilon(x^- - y^-)\delta^2(x_\perp - y_\perp),\end{aligned} \tag{2.25}$$

where the relation $\partial_x\varepsilon(x) = 2\delta(x)$ between the sign and delta distribution has been used. Therefore the basic Poisson brackets are

$$\left\{\phi(x^-, x_\perp), \phi(y^-, y_\perp)\right\}_{x^+=y^+} = -\frac{1}{4}\varepsilon(x^- - y^-)\delta^2(x_\perp - y_\perp). \tag{2.26}$$

Only the fields are involved in (2.26) because we have reduced the symplectic variables by the constraints (2.15) from the beginning. Using the brackets (2.26) in (2.17) one gets the correct equation of motion (2.19). Quantization is done by correspondence principle and leads to the field commutator

$$[\phi(x), \phi(y)]_{x^+=y^+} = -\frac{i}{4}\varepsilon(x^- - y^-)\delta^2(x_\perp - y_\perp). \tag{2.27}$$

This reduction of variables is interrelated to the question of initial data to the light front equation of motion which are first-order in LC time. The variational derivation starting from the action uses boundary conditions in space and time direction to label the classical trajectories of the field. For a second-order Lagrangian it is equivalent to label the classical trajectories by the initial data $\phi(x)|_{x^0=0}$ and the first time derivative $\frac{\partial\phi}{\partial x^0}|_{x^0=0}$. This initial data translates in the Hamiltonian formulation into data for the field $\phi(x)$ and the canonical momentum $\pi(x)$. In the front form it seems that initial data for the field at $x^+ = 0$ alone is enough. But by inspecting the equation of motion (2.19), one notices that inverting the x^--derivative requires additional initial data. Two possible ways are suggested: firstly, providing initial data on the second light front surface $x^- = 0$ makes the solution of (2.19) unique, but leads to a two-Hamiltonian formulation [13] and is thus unfavored. Secondly, we can explicitly demand some spatial boundary conditions in x^--direction for all light cone times. By using the relation $\partial_x \varepsilon(x) = 2\delta(x)$, where $\varepsilon(x)$ and $\delta(x)$ are distributions on some test function space, we have already implicitly obeyed some boundary conditions. In case of quantizing the theory in a finite volume other boundary conditions, like periodic ones to avoid surface terms, are usually fixed and one has to carefully determine the solution $G(x,y)$ of

$$\frac{\partial}{\partial x^-} G(x,y) = \delta(x^- - y^-), \tag{2.28}$$

since it influences the quantization. We will come back to this point in the Section 2.3 about discrete light cone quantization and write down the field commutator in finite volume quantization. The issue of inverting the operator ∂_{x^-} is also the reason for the constrained zero mode problem. If ∂_{x^-}, being an operator acting on some function space, has zero modes it is not invertible and consequently the inverse of the symplectic form ω^{-1} is not defined. Faddeev and Jackiw suggested to identify the zero modes of the 'matrix' $\omega(x,y)$ and use only an invertible subblock of ω to define the Poisson brackets, that is to leave the degrees of freedom related to the zero modes aside. For the LC case (2.23) the task would be to identify the zero mode contributions and subtract them accordingly, i.e. define a function space of fields where $G(x,y)$ in (2.28) is well-defined.

Fermions

Another subtlety encountered in light front quantization is the reduction from four to two independent components of Dirac spinors degrees of freedom. The free Dirac Lagrangian in light front coordinates is

$$\mathcal{L} = \bar{\Psi}\left(i\partial\!\!\!/ - m\right)\Psi, \tag{2.29}$$

where the partial derivative reads

$$\partial\!\!\!/ = \gamma_+\partial^+ + \gamma_-\partial^- + \gamma_i\partial^i. \tag{2.30}$$

The gamma matrices are $\gamma_\pm = \gamma_0 \pm \gamma_3$ analog to (2.1). It is possible to define the projectors $\Lambda^\pm = \frac{1}{4}\gamma^\mp\gamma^\pm$ and decompose the spinor into

$$\Psi = \Lambda^+\Psi + \Lambda^-\Psi = \Psi_+ + \Psi_- = \begin{pmatrix}\psi\\ \chi\end{pmatrix}, \tag{2.31}$$

where the explicit choice of the gamma matrices is explained in Appendix A. Inserting (2.31) in the Dirac Lagrangian (2.29) and eliminating all gamma matrices leads to

$$\mathcal{L} = 2i\left(\chi^\dagger\partial_-\chi + \psi^\dagger\partial_+\psi\right) - \psi^\dagger\sigma^j\partial_j\chi + \chi^\dagger\sigma^j\partial_j\psi - m\left(\psi^\dagger\chi + \chi^\dagger\psi\right). \tag{2.32}$$

One observes that there is no term in (2.32) containing a x^+-derivative of χ. Hence we call χ the non-dynamical and ψ the dynamical spinor component. The Euler-Lagrange equations for ψ and χ read

$$\frac{\partial\mathcal{L}}{\partial\psi^\dagger} = 2i\partial_+\psi \quad = (-\sigma^j\partial_j + m)\,\chi, \tag{2.33}$$

$$\frac{\partial\mathcal{L}}{\partial\chi^\dagger} = 2i\partial_-\chi \quad = (\sigma^j\partial_j + m)\,\psi, \tag{2.34}$$

here the latter equation poses again a constraint. One could run the same procedure via the symplectic two-form, that has lead to (2.26), to eliminate the non-dynamical spinor components χ and its canonical momentum variable, but instead we suppose that we have formally solved (2.34) and replaced χ in (2.33). We are then left with the canonical quantization of ψ, which has the independent canonical momentum

$$\Pi_\psi = 2i\psi^\dagger \tag{2.35}$$

and hence by canonical prescription the anti-commutation relations follow

$$\begin{aligned}\left\{\psi_\alpha(x),\psi^\dagger_\beta(y)\right\}_{x^+=y^+} &= \frac{1}{2}\delta_{\alpha\beta}\delta(x^- - y^-)\delta^2(x_\perp - y_\perp),\\ \left\{\psi_\alpha(x),\psi_\beta(y)\right\}_{x^+=y^+} &= \left\{\psi^\dagger_\alpha(x),\psi^\dagger_\beta(y)\right\}_{x^+=y^+} = 0,\end{aligned} \tag{2.36}$$

where $\alpha,\beta\in\{1,2\}$ are spinor indices. The crucial step above is again the inversion of ∂_- in (2.34), which poses the same problem as before for the scalar field. Ignoring the fermionic zero mode means to choose the Green function to be $G(x-y)=\frac{1}{2}\varepsilon(x-y)$ in (2.28), which works fine for free fermion fields.

Finally, we give the mode expansion of the dynamical fermion field since this representation is subsequently used. The solution of the Dirac equation following from (2.29) can be written in the Heisenberg picture as

$$\Psi(x) = \sum_s \int\limits_{k_->0} \frac{dk_- d^2k_\perp}{\sqrt{2k_-(2\pi)^3}} \left(b_s(k)u_s(k)e^{-ik_{\text{on}}\cdot x} + d^\dagger_s(k)v_s(k)e^{ik_{\text{on}}\cdot x}\right), \tag{2.37}$$

where k_{on} stands for the on-shell four-momentum. Using the gamma matrix notation given in Appendix A the elementary spinors in LC coordinates can be represented as

$$\begin{aligned} v_1(k) &= \left(\sqrt{2k_-}, 0, -\frac{m}{\sqrt{2k_-}}, \frac{ik_1 - k_2}{\sqrt{2k_-}}\right)^T, \quad v_2(k) = \left(0, \sqrt{2k_-}, \frac{ik_1 + k_2}{\sqrt{2k_-}}, -\frac{m}{\sqrt{2k_-}}\right)^T,\\ u_1(k) &= \left(\sqrt{2k_-}, 0, \frac{m}{\sqrt{2k_-}}, \frac{ik_1 - k_2}{\sqrt{2k_-}}\right)^T, \quad u_2(k) = \left(0, \sqrt{2k_-}, \frac{ik_1 + k_2}{\sqrt{2k_-}}, -\frac{m}{\sqrt{2k_-}}\right)^T. \end{aligned} \tag{2.38}$$

The regular normalization properties apply

$$u_s^\dagger(k)u_{s'}(k) = v_s^\dagger(k)v_{s'}(k) = \frac{1}{2}(k_+ + k_-)\delta_{ss'} = k_0\delta_{ss'}. \tag{2.39}$$

Furthermore one has the familiar completeness relations of $u_s(p)$ and $v_s(p)$, namely

$$\begin{aligned}\sum_s u_s(k)\bar{u}_s(k) &= \not{k} + m,\\ \sum_s v_s(k)\bar{v}_s(k) &= \not{k} - m.\end{aligned} \tag{2.40}$$

The spinors in (2.38) are not independent because of (2.34). Projecting with the help of $\Lambda^\pm$ one introduces $u_s^\pm = \Lambda^\pm u$ and analogously for v_s. These are given in the notation of Appendix A by the upper (lower) two components of u (v) in (2.38), as has been pointed out above in (2.31). Hence the dynamical spinor components simply read

$$\psi(x) = \int_0^\infty \frac{dk_-}{\sqrt{(2\pi)^3}}\frac{1}{\sqrt{2k_-}}\sum_s \int d^2k_\perp \left[u_s^+(k)e^{-ik_{\text{on}}\cdot x}b_s(k) + v_s^+(k)e^{ik_{\text{on}}\cdot x}d_s^\dagger(k)\right]. \tag{2.41}$$

Normalizations of u_+, v_+ are then

$$\frac{1}{2}u_{+s}^\dagger(k)u_{+s'}(k) = \frac{1}{2}v_{+s}^\dagger(k)v_{+s'}(k) = k_-\delta_{ss'}. \tag{2.42}$$

The quantization rules (2.36) imply the commutator algebra of the mode creation and annihilation operators

$$\begin{aligned}\left\{b^\dagger(p), b(q)\right\} &= \delta(p^- - q^-)\delta^2(p_\perp - q_\perp),\\ \left\{d^\dagger(p), d(q)\right\} &= \delta(p^- - q^-)\delta^2(p_\perp - q_\perp),\end{aligned} \tag{2.43}$$

and all other combinations vanish.

Gauge theory

The SU(N) gauge theory in light cone gauge ($\alpha \to 0$) is given by the Lagrangian

$$\mathcal{L} = \text{Tr}\, G_{\mu\nu}G^{\mu\nu} - \frac{1}{2\alpha}(n_\mu A_a^\mu)^2, \tag{2.44}$$

where the gauge field strength tensor is the commutator of the covariant derivatives $G^{\mu\nu} = [D^\mu, D^\nu]$ and the constant four-vector n is light-like. The gauge fields $A_\mu(x) = \sum_a A_\mu^a(x)T_a$ are elements of the Lie algebra and the generators T_a obey the SU(N) algebra

$$[T_a, T_b] = f_{ab}{}^c T_c, \tag{2.45}$$

where ${f_{ab}}^c$ are the structure constants. The gauge fixing term and the limit $\alpha \to 0$ are equivalent to the gauge condition

$$n^\mu A^a_\mu = 0. \tag{2.46}$$

Light cone gauge belongs to the family of non-covariant axial gauges. We refer the reader to the review article [27] on axial gauges and in particular light cone gauge. One distinguishes three subsets in non-covariant gauges depending whether the four-vector n is space-like (axial gauges, planar gauge), time-like (temporal gauge) or light-like. The axial gauge and the light front gauge have the desirable feature of the decoupling of the ghost fields in common. This is easily seen by inspecting the ghost term for the Yang-Mills Lagrangian (2.44)

$$\mathcal{L}_{gh} = \bar{\eta}^a n^\mu D^{ab}_\mu \eta^b, \tag{2.47}$$

where $n^\mu D^{ab}_\mu = \delta^{ab} n^\mu \partial_\mu + g f^{abc} n^\mu A^c_\mu$. The ghost-gluon vertex and the ghost propagator are proportional to n, thus any diagram involving ghosts and gluons (or in more general theories all physical fields) vanishes and the ghosts decouple. Even more directly, the coupling term in (2.47) is absent when the gauge condition (2.46) is used. The advantage is that the theory is solely formulated in physical degrees of freedom from the beginning. No negative norm states in Hilbert space have to be introduced and the unitarity of the S-matrix is not questioned at any stage. However, the light cone gauge fixing is not Lorentz invariant and therefore the covariance of the equations is not manifest. In principle, the covariance has to be checked explicitly. Moreover, perturbative computations are complicated by non-covariant terms in the gluon propagator absent in covariant gauges. Finding counterterms in perturbative renormalization also is more difficult than in non-covariant gauges because the set of allowed operators is not restricted by Lorentz symmetry.

The setup of light front gauge is justified even without addressing the framework of light cone quantization. But when light front quantization is applied the most natural gauge choice is (2.46). Fixing $n_\mu = (1, 0, 0, 0)$ reduces the gauge condition to

$$n \cdot A_a = A^+_a = 0. \tag{2.48}$$

Gauging away the $+$ component of the gauge field strongly resembles the temporal gauge in the instant form and one expects some of its features to be present. Light cone gauge does not fix the gauge freedom completely, because gauge transformations independent of x^+ are conform with the gauge condition (2.48). The residual gauge freedom is erased by solving the non-dynamical Gauss law

$$\partial^2_- A^-_a = J^+_a, \tag{2.49}$$

where a conserved current was added on the r.h.s., usually given by combinations of matter fields and the transverse gauge fields. Equation (2.49) is a constraint equation for the gauge fields A^-_a as one finds in instant form temporal gauge. In contrast to the experience in temporal gauge where the Gauss law operator is in general non-linear (an important exception is the abelian gauge theory), the constraint above is a linear equation for A^- also for non-abelian gauge theories. Consequently, the classical formal solution of (2.49) can be used to substitute

A^- everywhere and thereby only the transverse components of the gauge field remain. In general, canonical quantization should be consistently done by application of the Dirac-Bergman or Faddeev-Jackiw formalisms as mentioned in the paragraph on scalar fields. Like for the fermions here we only state the final commutation relations

$$\left[A_i^a(x^-, x_\perp), A_j^b(y^-, y_\perp)\right]_{x^+=y^+} = \frac{i}{4}\varepsilon(x^- - y^-)\delta(x_\perp - y_\perp)\delta_{ij}\delta_{ab}, \quad i, j = 1, 2. \tag{2.50}$$

Again putting forward the free field expansion of the gauge fields

$$A_\mu^a(x) = \sum_\lambda \int \frac{dp^+ d^2 p_\perp}{\sqrt{2p^+(2\pi)^3}} \left(a_{\lambda,a}(p)\varepsilon_\mu(p,\lambda)e^{-ip_{\text{on}}x} + a_{\lambda,a}^\dagger(p)\bar{\varepsilon}_\mu(p,\lambda)e^{ip_{\text{on}}x} \right) \tag{2.51}$$

the commutators (2.50) are equivalent to

$$\left[a_{\lambda',a}(p), a_{\lambda,b}^\dagger(q)\right]_{x^+=y^+} = \delta(p^+ - q^+)\delta(p_\perp - q_\perp)\delta_{\lambda\lambda'}\delta_{ab}, \tag{2.52}$$

with all other combinations vanishing.

2.3. Discrete light front quantization

A convenient infrared regularization of light front field theory in 1+1 dimensions is given by discrete light cone quantization (DLCQ). Here we consider only two-dimensional theories and review the approaches to generalize DLCQ to four-dimensional theories in the second part of this section. In DLCQ the physical system is placed inside a light-like box

$$-\frac{L}{2} \leq x^- \leq \frac{L}{2}. \tag{2.53}$$

Furthermore one demands periodic or anti-periodic boundary conditions for the fields. The first option is equivalent to the compactification of the x^- direction to a circle of length L. Periodic boundary conditions are essential for the conserved currents of the theory, since otherwise the conservation law would be violated by surface terms. DLCQ has been first introduced in [9, 28] and consequently applied to a number of models, some examples are given in [29, 30, 31, 32]. The boundary conditions lead to discretized momentum modes of the fields

$$k_n^+ = \frac{2\pi}{L}n, \quad n \in \mathbb{N}, \tag{2.54}$$

and the field theory can be formulated as quantum many-body theory. There is the freedom to choose different basis functions in the one-particle Hilbert space. Usually one uses the free particle expansion, that is plane waves, as basis functions. Other choices like oscillator wavefunctions are possible and should be made depending on the symmetries and other constraints on the physical system.

In the following we will be interested in the spectrum and eigenstates of the theory and therefore we express the momentum operators in the DLCQ basis. The explicit form of these operators depends of course on the theory under consideration and the light cone momentum operators for the Schwinger model are given in Section 4.2. Here we emphasize the generic structure of these operators and give the expressions for real scalar fields in two dimensions. The mode expansion of the scalar field at $x^+ = 0$ reads

$$\phi(x) = \frac{1}{L}\sum_n \frac{1}{\sqrt{n}}\left(a_n e^{-ik_n^+ x^-} + a_n^\dagger e^{ik_n^+ x^-}\right). \tag{2.55}$$

Inserting the mode expansion (2.55) into the kinematical operator P^+ one derives

$$P^+ = \frac{2\pi}{L}\sum_n n a_n^\dagger a_n, \tag{2.56}$$

which is diagonal in the DLCQ basis. Hence the LC Hamiltonian becomes

$$P^- = \frac{L}{2\pi}\left(\sum_n \frac{m^2}{n} a_n^\dagger a_n + V\right), \tag{2.57}$$

where V is an unspecified potential. With the box size L one defines

$$\begin{aligned} K &= \frac{L}{2\pi} P^+, \\ H_{\mathrm{LC}} &= \frac{2\pi}{L} P^-, \end{aligned} \tag{2.58}$$

wherein the dimensionless K is called harmonic resolution and measures the approximation toward the continuum theory. The LC mass operator H_{LC} has dimension mass squared $[m^2]$. For practical purpose a second, ultraviolet cut-off is needed which restricts the number of modes, i.e. $n \leq \Lambda$. This cut-off is equivalent to a maximum resolution K_{max} used in the computation. Since the considered interactions are P^+ conserving, the Fock space can be divided into sectors of constant resolution and the LC Hamiltonian can be treated sector by sector. Finally the results obtained at finite resolution are extrapolated to $K \to \infty$ which is equivalent to the infinite volume limit $L \to \infty$ at a fixed momentum P^+.

Some concerns about microcausality violation in the finite volume formulation have been reported in the simple free scalar LF field theory, see [20]. It was proposed to give up Lorentz invariance as result of the observation that LF field theory is incompatible with the Wightman axioms [33]. However, these concerns have been overcome for the scalar models [34, 35]. The conclusion is that at finite volumes L the causality structure of the theory is violated but this violation is suppressed by powers of L and therefore vanishes in the continuum limit.

Essentially, there are two avenues to take DLCQ to higher dimensions. The first is to banish the system also into a box in the two spatial transverse directions $x_\perp$ in addition to the L-box in x^- [36, 37]. Of course the discrete transverse momenta are indefinite in sign which implies a huge enhancement of basis states.

Nevertheless, the problem of solving a quantum field theory is 'reduced' to dealing with coupled many-body equations,

$$\left[M^2 - \sum_{i=1}^{n} \frac{k_{\perp i}^2 + m_i^2}{x_i}\right] \psi_n = \sum_{n'} \int \langle n|V_{\rm LC}|n'\rangle \psi_{n'}, \tag{2.59}$$

where ψ_n is the projection of $|\Psi\rangle$ on the n-particle Fock state $|n\rangle$. Besides the maximal harmonic resolution K an additional cutoff, preferably gauge and Lorentz invariant, is needed to restrict the transverse constituent momenta. One suggestion is the Brodsky-Lepage cutoff [36]

$$\sum_i \frac{k_{\perp i}^2 + m_i^2}{k_i^+} \leq \Lambda^2 \tag{2.60}$$

as a covariant ultraviolet regularization. This truncation removes Fock states which are far off the energy shell of the bound state since

$$\begin{aligned} \sum_i p_i^- - P_i^- &= \sum_i \left(\frac{(k_{\perp i}^2 + x_i P_\perp)^2 + m_i^2}{x_i P^+}\right) - \frac{P_\perp^2 + M^2}{P^+} \\ &= \frac{1}{P^+}\left(\sum_i \frac{k_{\perp i}^2 + m_i^2}{x_i} - M^2\right), \end{aligned} \tag{2.61}$$

where $(P^+, M, P_\perp)$ denote the bound state four-momentum and $k_{\perp i}$, $x_i = k_i^+/P^+$ the relative momenta connected to the physical momenta of the constituents by $p_{\perp i} = k_{\perp i} + x_i P_\perp$. From (2.59) one observes that the common structure of LF wavefunctions is formally

$$\psi_n = \frac{\Gamma_n}{M^2 - \sum\limits_{i=1}^{n} \frac{m_i^2 + k_{\perp i}^2}{x_i}}, \tag{2.62}$$

where the vertex function Γ_n is given by $\Gamma_n = \sum_{n'} \int V_{nn'} \psi_{n'}$. Contributions to the LC wavefunction from far-off-shell Fock states are expected to be suppressed. In contrast to the Tamm-Dancoff approximation no cut on particle number is employed. However, a further restriction of particle number may be necessary from the computational point of view. If such a cut is applied special care has to be paid to retain gauge invariance within the Fock subspace.

The DLCQ matrix equation is numerically the trapezoid approximation to the set of integral equations coupling the various Fock sectors. Improvements of convergence are gained by advantageous sampling of the integrand instead of equidistant points and by using different orthogonal basis functions instead of plane waves, as well as using other techniques common in computational nuclear and atomic physics. Results were reported on the Positronium spectrum of three-dimensional QED at the strong coupling $\alpha = 0.3$, where additionally a Tamm-Dancoff restriction to the three-particle Fock space was applied [37].

Another idea is to combine the DLCQ matrix formalism with a transverse space lattice. This was very early put forward in [38] and later on investigated for example in [39, 40, 41]. See [12] for a review of recent progress. In detail, one splits space-time into two continuous directions x^+, x^- and a lattice $x_\perp + a\hat{r}$, where a is the lattice spacing and $\hat{r} = (n_1, n_2)$ with $n_i \in \mathbb{N}$ is a generic lattice point in the transverse plane. At each lattice point $x_\perp$ one introduces SU(N) link matrices $M_r(x_\perp)$ directing from $x_\perp$ to $x_\perp + a\hat{r}$. Additionally, the continuous gauge fields $A_\alpha(x_\perp, x^+, x^-)$ with $\alpha = +, -$ are defined at each lattice point $x_\perp$. The gauge field and the link matrices transform under transverse gauge transformations as follows

$$\begin{aligned} A_\alpha(x) &\to U(x)A_\alpha(x)U^\dagger(x) + (\partial_\alpha U(x))U(x), \\ M_r(x_\perp) &\to U(x_\perp)M_r(x_\perp)U^\dagger(x_\perp + r). \end{aligned} \tag{2.63}$$

As usual the pure gauge lattice action is found by considering the simplest gauge invariant (here only under transverse gauge transformations) action, which reduces to the classical action in the naive continuum limit. Taking into account only the transverse plaquette the transverse lattice action reads

$$\begin{aligned} S_{tl} = \sum_{x_\perp} \int dx^+ dx^- \Bigg(&-\frac{a^2}{2g^2} \mathrm{Tr}\left\{F^{\alpha\beta}F_{\alpha\beta}\right\} + \frac{1}{g'^2} \sum_r \mathrm{Tr}\left\{D_\alpha M_r(x_\perp)\left(D^\alpha M_r(x_\perp)\right)^\dagger\right\} \\ &+ \frac{1}{g''a^2} \sum_p \mathrm{Tr}\left(\mathrm{Re}\, U_p(x_\perp)\right) \Bigg), \end{aligned} \tag{2.64}$$

where $U_p = U_{rs}$ is the elementary transverse plaquette with the two directions r, s and the sum over p covers the whole transverse plane. Explicitly, the transverse plaquette reads

$$U_p = U_p^\dagger = U_{rs}(x_\perp) = M_r(x_\perp)M_s(x_\perp + r)M_r^\dagger(x_\perp + s)M_s^\dagger(x_\perp). \tag{2.65}$$

The covariant derivative is given by

$$D_\alpha M_r(x_\perp, x^+, x^-) = \partial_\alpha M_r + iA_\alpha(x_\perp)M_r - iM_r A_\alpha(x_\perp + r). \tag{2.66}$$

We outline the general strategy in the following. Demanding a gauge condition (typically light cone gauge $A^+ = 0$) and solving formally the classical constraint equation for A^- one eliminates $A_\alpha(x)$ completely in (2.64). The light cone Hamiltonian and the harmonic resolution are derived. It is pointed out in [38, 39] that the action (2.64) contains the non-linear gauged SU(N) sigma model in the M_r variables and ideally the solution to the free SU(N) non-linear sigma model would be used as basis states to construct solutions of the full theory. Since the solution to the non-linear sigma model is not known one has followed a more direct approach in the literature replacing the SU(N) link matrices by general linear matrices and adding a potential to (2.64) which ensures the matrices M_r to be SU(N) in the continuum limit. Having $M_r(x_\perp) \in$ GL(N) the sigma model is now linear and the ground state is $M_r = 0$. The link matrices are canonically quantized at equal light cone time and are represented at $x^+ = 0$ in Fock space as

$$M_{r,ij}(x^-, x_\perp) = \frac{1}{\sqrt{4\pi}} \int \frac{dk^+}{\sqrt{k^+}} \left(a_{-r,ij}(k^+, x_\perp) e^{-k^+ x^-} + a^\dagger_{r,ij}(k^+, x_\perp) e^{ik^+ x^-} \right), \quad (2.67)$$

where $a^\dagger_{r,ij}(k^+, x_\perp)$ creates a link parton with momentum k^+ transporting color i from $x_\perp$ to j at $x_\perp + r$. Using DLCQ in the x^- direction the longitudinal momenta k^+ are discretized as in the two-dimensional case. Finally, the basis states are constructed out of link partons and the Hamiltonian matrix is found and diagonalized to obtain the bound state spectrum and the LC wavefunctions. The several couplings present in (2.64) or even more in improved versions of the transverse lattice action have to be tuned such that the theory is critical at finite lattice spacing to give unambiguous answers. One way of realizing this is to look for regions in the space of coupling constants where enhanced symmetry properties like Poincaré, chiral symmetry happen to be present, since these are expected to be restored in the continuum limit. So far only two and four link states have been considered but even this crude approximation has lead to many promising results like glue ball masses and the pion LC wavefunction in good agreement with four-dimensional lattice gauge theory.

2.4. The zero mode challenge

This section will sketch one of the conceptual challenges in the light front framework, the infamous zero mode problem. We have mentioned this issue in Section 2.2 where the inversion of the derivative operator was necessary. Supposedly zero modes, which are the only excitations kinematically allowed in the LF vacuum, are crucial to describe spontaneous symmetry breaking and other vacuum phenomena in LF quantization. Regarding the spontaneous breaking in simple models, like the Z_2 symmetry $\phi \leftrightarrow -\phi$ in scalar ϕ^4 in 1+1 dimensions, different claims were made in the literature, see [42, 43, 44, 45, 46, 26, 47, 48] for the necessity of the zero mode in the LF description and [49, 50] against it. One must note, that the latter authors usually performed simulations in the symmetry broken phase of ϕ^4_{1+1} (for negative mass) and thereby are able to observe the degeneracy of the lowest states in the particle number even and odd sectors. With periodic boundary conditions degeneracy of particle odd and even sectors is equivalent to the restoration of the parity symmetry. For a DLCQ application to ϕ^4_{1+1} in the broken phase omitting the zero mode, see [51, 52, 53]. Here we will not discuss the arguments leading the authors to their respective conclusions in detail, but instead simply review the different sorts of zero modes commonly encountered in LF quantized field theories and the difficulties they present.

The local zero mode is the x^- independent part of the field $\phi(x)$, i.e.

$$\phi_0(x^+, x_\perp) = \frac{1}{L} \int_{-L/2}^{L/2} dx^- \phi(x). \quad (2.68)$$

It is common to define the normal mode field as

$$\phi_n(x) = \phi(x) - \phi_0(x). \tag{2.69}$$

Further the transverse zero modes ϕ_0^i, $i = 1, 2$ are given by integration of ϕ_0 with respect to the transverse direction. Finally, one defines the global zero mode $\overset{0}{\phi} = \int d^2x_\perp \phi_0$. The zero mode degrees of freedom usually become important when the theory is treated on a (higher-dimensional) torus and non-trivial boundary conditions are demanded for the field. A clear separation of zero mode and normal modes is only possible in a finite volume quantization. In the LF framework the $k^+ = 0$ mode is an accumulation point of the spectrum and so no clear cut can be made in the continuum.

In two dimensional models the local zero mode is identical to the global one. The definition (2.68) is equivalent to a projection on the orthogonal subspace spanned by the plane wave solutions $e^{ip^- x^+}$ with $p^- = 0$. All equations may be projected to the zero mode and normal mode sectors. In case the theory is linear the zero mode and normal mode sector will decouple. However, for non-linear scalar theories one finds that the zero and normal mode sector are coupled due to the zero mode constraint

$$-m^2\phi_0 + \frac{1}{L}\int_{-L/2}^{L/2} dx^- \, \frac{\partial V(\phi)}{\partial \phi} = 0. \tag{2.70}$$

Accordingly, the zero mode is not an independent field degree of freedom and should be eliminated before the theory is quantized. Equation (2.70) arises from the projection to the zero mode sector of the equation of motion. Alternatively, the zero mode constraint is found in the Dirac-Bergman algorithm as a consistency condition of the canonical momentum constraints. Treating Eq. (2.70) as an operator equation one has to carefully specify the operator ordering to make this equation unambiguous. The constraint is so far only solved by perturbation theory, Tamm-Dancoff approximation to one- and two-particle Fock states and mean field approximation, see [13, 26]. To circumvent these operator ordering issues one may just find the classical solutions to determine the effective potential or follow some variational ansatz.

Anyway, one collects corrections coming from (2.70) into a effective potential familiar from instant form discussions of spontaneous symmetry breaking. Replacing the zero modes in the Hamiltonian by different (perturbative) solutions of (2.70) leads to various Hamiltonians describing the symmetric phase as well as the symmetry broken phase. On the contrary to the instant form phenomenology the vacuum state stays unchanged and trivial, since the zero mode quanta are subtracted from the theory.

The different LC Hamiltonians in the broken and symmetric phases are analogous to the alteration of the instant form Hamiltonians by shifting the relevant fields by their vacuum expectation values (VEV). The VEV of the shifted field is of course zero and explicitly present in the transformed Hamiltonian.

Similar considerations can be made in fermionic theories, especially devoted to the question of chiral symmetry breaking. As mentioned in Section 2.2 in LF quantization only half of the spinor component fields are dynamical. In the solution of the constraint (2.34) the zero mode of Ψ_- is conventionally discarded by choosing anti-periodic boundary conditions. This is admissible because the fermion fields show up only in bilinears. As a consequence even the free, massive Dirac field is chirally symmetric on the light front. Whether the LC chiral transformation, which is only defined on the dynamical spinor components, provides equivalent physical informations as the common transformation is unclear.

Most important for the following chapters will be the occurrence of the zero modes of gauge fields. We will only introduce the zero mode of the light cone components of the gauge field since we restrict ourselves to two-dimensional models. Like for the scalar fields the zero mode of $A^{\pm}(x)$ is

$$A_0^{\pm}(x^+) = \frac{1}{L} \int_{-L/2}^{L/2} dx^- A^{\pm}(x). \tag{2.71}$$

The zero mode A_0^+ is gauge-invariant under gauge transformations connected to the identity if periodic boundary conditions in x^- are implemented. Therefore A_0^+ can not be gauged away and presents a dynamical degree of freedom in contrast to the scalar models. But by gauge fixing one can ensure that the zero mode A_0^- and the normal modes A_n^+ vanish. Yet the normal modes of A^- are constrained by the Gauss law

$$\partial_-^2 A_n^- = \mathcal{F}(\psi)\,, \tag{2.72}$$

where $\mathcal{F}(\psi)$ is some function of the matter fields coupled to the gauge fields $A^{\pm}$. Thereby the only remainder after gauge fixing is the zero mode A_0^+ which obeys some equation of motion and is a quantum mechanical background field. Topological effects found in gauge theories should in the LF formulation be related to the dynamics of the zero mode and the structure of the corresponding Fock sector. In Section 4.2 the gauge field zero mode of the $U(1)$ gauge field in the Schwinger model will be treated consistently and be responsible for the degenerate vacuum states, called theta vacua.

In conclusion, zero modes of the fundamental fields are the only kinematical allowed longitudinal excitations in the ground state of a LF quantized field theory. In this sense the vacuum structure on the LF is simplified compared to the instant form but maybe not trivial. Whether zero modes are important depends on the boundary conditions specified. Additionally, the zero modes can be independent or constrained degrees of freedom depending on the theory under consideration. Using the solution of the constraint equation the constrained zero modes can be integrated out and the triviality of the LF ground state is established.

3. Relativistic statistics and thermodynamics

In this chapter we treat finite temperature field theory in the framework of light cone quantization. A proper relativistic generalization of the statistical physics and thermodynamics is used to relate LC thermal field theory to common phenomenological quantities. First attempts tried to utilize the LC Hamiltonian P^- to formulate the theory, but failed to obtain equivalence to results known from instant form consideration.

We start with relativistic statistical physics which involves the four-velocity u of the medium [54, 55] and project the invariant expressions back to the light front frame. One achieves a representation where the LF statistical operator is given by its instant form analog expressed in LF fields [56, 57]. It is possible to introduce so-called generalized LF coordinates which are natural to use for the LF thermal field theory [58, 8, 59].

One may argue that in doing so one actually does not perform LC quantization in the strict sense of Section 2. However, the method outlined ensures the equivalence of the light-front and instant form finite temperature results. We turn to this specific question in the last paragraph where it is shown that the (standard) definition of LC temperature of a non-interaction gas in the micro-canonical ensemble appears to be inconsistent.

Some applications of LF thermal field theory are given in [60, 61, 62, 63, 64, 65, 66]. As a perspective the framework should be very useful for the physics of heavy ion collisions since there one is interested in (boost-invariant) properties of hadrons at high temperature and density. Attempts to define medium-dependent parton distributions have been reported in [56].

3.1. Covariant statistical mechanics

We follow the general derivation of the quantum statistical operator ϱ, see [67] in a general context and LF specific [56], representing a mixed state

$$\varrho = \sum_{n,m} c_{nm} |\psi_n\rangle\langle\psi_m|. \tag{3.1}$$

From the light cone Schrödinger equation for the states $i\partial_+|\psi_n\rangle = P^-|\psi_n\rangle$ one obtains the von Neumann equation

$$i\partial^- \varrho = [P^-, \varrho], \tag{3.2}$$

which holds for closed statistical systems. In equilibrium the right hand side of (3.2) vanishes and ϱ commutes with the Hamiltonian, hence is a constant of motion. We demand a decomposition property of ϱ describing the union of two statistical independent subsystems

$$\varrho = \varrho_1 \varrho_2, \tag{3.3}$$

where ϱ_1, ϱ_2 are the density matrices of two subsystems. Condition (3.3) implies the additivity of $\ln(\varrho) = \ln(\varrho_1) + \ln(\varrho_2)$. The statistical systems under consideration should be Poincaré symmetric and have further unspecified symmetries, e.g. charge or particle number conservation. By Noethers theorem symmetries come with conserved charges from which some are additive. The operator $\ln(\varrho)$ is a new additive charge not given by symmetry and is in general a linear combination of the other additive constants of motion. In the instant form one has seven integrals of motion related to the Poincaré symmetry, namely the three angular momenta J_i and four momenta P^μ. An additional charge due to Poincaré symmetry is the center of mass movement, which is not additive. One concludes that the general form of the statistical operator $\varrho_{\rm IF}$ (IF for instant form) is

$$\ln(\varrho_{\rm IF}) = \alpha + \beta_\mu P^\mu + \gamma_i J^i + \mu_a N^a. \tag{3.4}$$

Usually the translation and rotation symmetry is broken by a fixed box in the spatial directions to enclose the statistical system, thus only energy is left as constant of motion [67]. This brings $\varrho_{\rm IF}$ into the standard form

$$\ln(\varrho_{\rm IF}) = \alpha + \beta\left(P^0 - \mu_a N^a\right). \tag{3.5}$$

So far nothing has been said about statistical ensembles and the constants β, μ_a are totally unconstrained, so (3.4) should not been hasty identified as grand-canonical statistical operator or similar.

Analogous reasoning leads to a front form statistical operator having the structure of (3.4). But one realizes immediately from (2.8) that in the front form, by choosing x^+ as time evolution parameter, the boosts and angular momentum operators get mixed in the kinematical generators $\{M^{12}, M^{-1}, M^{-2}\}$. This forces the problem of how to treat rotations in LF theory, as discussed in [10] and references therein, on our attempt. We shall not enter these developments and ignore this issue by addressing only non-rotating subsystems or break LF rotational symmetry in some way from here on. In conclusion of the arguments above we make the linear ansatz

$$\ln(\varrho_{\rm LF}) = \alpha + \beta_\mu P^\mu + \mu_a N^a, \tag{3.6}$$

for the LF statistical operator where α is fixed by normalization and β^μ, μ_a are arbitrary constants.

The structure of the statistical operator in (3.6) is utilized in covariant statistical mechanics and thermodynamics where the heat bath is moving with four-velocity u_μ [68, 54, 55, 69]. There the four-vector β_μ is expressed as

$$\beta_\mu = \frac{1}{T} u_\mu, \tag{3.7}$$

where T is the rest frame temperature (proper temperature). The basic quantities of covariant thermodynamics characterizing the state of the statistical system are the entropy flux density s^μ, the energy momentum tensor $T^{\mu\nu}$ and other current densities n_a^μ, e.g. particle number, charge or mass. The usual local conservation laws hold

$$\begin{aligned} \text{energy-momentum:} \quad & \partial_\nu T^{\mu\nu} = 0, \quad T^{\mu\nu} = T^{\nu\mu}, \\ \text{other conserved quantities:} \quad & \partial_\mu n_a^\mu = 0, \\ \text{entropy:} \quad & \partial_\mu s^\mu = \sigma \geq 0, \end{aligned} \tag{3.8}$$

where the last inequality is the second law of thermodynamics and σ summarizes all entropy source terms. Let us assume that the thermodynamical system is in a state of local equilibrium. One can image the system being divided into subsystems each large enough to be a thermodynamical system in equilibrium but small against the overall size. Local equilibrium is reached when each subsystem is the equilibrium but processes can still happen between different subsystems. By introduction of a four-velocity field one assigns each subsystem a trajectory and the whole system is the union of these world lines. This setup corresponds to the Eckart frame. Other definitions for the velocity field like the Landau frame are possible, but not important here. In each comoving frame Σ_a the standard non-relativistic equilibrium thermodynamical laws hold. It is useful to define the projector

$$h^{\mu\nu} = g^{\mu\nu} - u^\mu u^\nu, \tag{3.9}$$

which helps to decompose every tensor quantity into parts parallel and orthogonal to u_μ. The important thermodynamical quantities are lifted to four-vectors as

$$s^\mu = s u^\mu, \tag{3.10}$$

$$n_a^\mu = n_a u^\mu, \tag{3.11}$$

$$T^{\mu\nu} = e u^\mu u^\nu - p\, h^{\mu\nu}. \tag{3.12}$$

Here p,e,s and n_a are the invariant pressure, energy density, entropy density and the densities n_a according to the currents N_a. The first law can be relativistically generalized to

$$\beta_\nu dT^{\nu\mu} = ds^\mu + \sum_a \alpha_a dn_a^\mu \tag{3.13}$$

and is in the comoving frame (projecting (3.13) in direction of u^μ) equal to the standard one. Covariant grand-canonical statistical mechanics provides a statistical operator of the form (3.6) which leads to covariant thermodynamics from the definitions of entropy (density)

$$\mathrm{Tr}\,\{\varrho \ln \varrho\} = S = \int d\omega_\mu s^\mu, \tag{3.14}$$

where the surface element is given by $d\omega_\mu = d^4x\delta(u\cdot x)u_\mu$ and the ensemble conditions

$$\begin{aligned} \text{four-momentum:} \quad & P^\mu = \langle \hat{P}^\mu \rangle = \int d\omega_\nu \, \langle \hat{T}^{\nu\mu} \rangle, \\ \text{charges:} \quad & Q_a = \langle \hat{N}_a \rangle = \int d\omega_\nu \, \langle \hat{n}_a^\nu \rangle. \end{aligned} \tag{3.15}$$

Exceptionally, we have denoted operators by hatted symbols to distinguish them from the averaged quantity but this notation is only used in Equation (3.15). It should be clear from context if operators or macroscopic quantities are meant. Since the Lorentz transformation properties of P^μ and n^μ should not be changed by the statistical averaging (3.15), the statistical operator is a Lorentz scalar. Comparison with the relativistic Gibbs-Duhem equation or the first law (3.13) leads to the identifications (3.7) for β_ν and $\mu_a = T\alpha_a$ for the chemical potentials. Optimizing the entropy (3.14) under the constraints (3.15) one obtains the grand-canonical covariant statistical operator

$$\varrho = \frac{1}{\mathcal{Z}} \exp\left[\int d\omega_\mu \left(-\beta_\nu T^{\nu\mu} - \sum_a \mu_a n_a\right)\right] = \frac{1}{\mathcal{Z}} \exp\left[-\frac{1}{T}\left(u_\nu P^\nu - \sum_a \mu_a N_a\right)\right]. \tag{3.16}$$

In the special case of no relative velocity between heat bath and system one can evaluate the product $u \cdot P$ in LC coordinates with $(u^+, u^-, u^\perp) = (1, 1, 0^\perp)$, i.e. the heat bath at rest in the instant form, and finds

$$\varrho = \frac{1}{\mathcal{Z}} \exp\left\{-\frac{1}{T}\left(\frac{P^+}{2} + \frac{P^-}{2} - \sum_a \mu_a N_a\right)\right\}, \tag{3.17}$$

which has been suggested in [70] for scalar thermal field theory and in [71] in the context of the NJL model. Choosing $u^+ = 1$ and $u^- = 0, u^i = 0$ is prohibited since the four-vector u has to be time-like. Some earlier unpublished work [72] observed the problems coming from the naive generalization and suggested a solution similar to (3.17) arguing with the NLC coordinates (2.12).

In conclusion, we deduced a statistical operator (3.17) suitable for a LC quantized system by using the covariant generalization of statistical mechanics and then projecting the covariant operator back to the light front. This reasoning is in some sense unsatisfactory because we explicitly referred to the instant form. The statistical operator given in (3.17) describes a system which only evolves in instant form imaginary time. In the next section a different way to derive (3.17) is outlined without mentioning the instant form. The idea is to enlarge the set of linear transformations beyond the LC transformation and restrict to those coordinate transformations which make a non-singular partition function possible. Especially the original LC transformation (2.1) is not allowed [8] and a set of coordinates is introduced which compromise between thermal and light cone properties. Different claims regarding the equivalence of LC quantized thermal field theory and the conventional theory have been made in the literature, see [60, 73, 74], but using general light cone coordinates the equivalence of both formulations is established for physical systems where the zero temperature equivalence has been shown.

3.2. General light front thermal field theory

In formulating thermal field theory within the different relativistic forms (instant and front form) one has to be cautious to distinguish between the velocity of the heat bath u_μ on the one hand and time direction vector n^ν, which defines the quantization plane, on the other. In the instant form this causes no harm because it is always possible to use $u \cdot x = 0$ as quantization plane and $u \cdot P$ as time evolution operator. Since u is time-like it is possible to boost to the comoving frame. The theories quantized at $x^0 = 0$ and $u \cdot x = 0$ are equivalent because of the intrinsic covariant formulation of quantum field theory. In the rest frame of the medium one has the standard partition function

$$\mathcal{Z} = \mathrm{Tr}\left\{\exp\left(-\frac{\beta}{\sqrt{g_{00}}}P^0 - \mu_a N_a\right)\right\}. \tag{3.18}$$

As pointed out in the last section, if the system is moving relative to the heat bath with velocity u but is quantized at equal x^0 we can use the density operator (3.16). This situation changes when we switch to the light front. No boost transformation is possible to get to the rest frame and use (3.18) in the comoving frame. In fact if one tries to utilize (3.16) in LC coordinates the statistical operator is ill-defined since the metric component g_{++} vanishes.

One may ask if sets of coordinates could be defined which change the LC coordinates only little and avoid the problem of the vanishing metric component g_{++}. Here we will follow Weldon and Das [8, 59, 57] and investigate general light cone (GLC) coordinates related to the Minkowski ones by a linear transformation and later on specialize these considerations to obtain (3.17). Utilizing the notation of [8] we introduce the general light cone frame

$$\begin{aligned} x^{\bar{0}} &= a x^0 + b x^3, \\ x^{\bar{3}} &= c x^0 + d x^3, \\ x_\perp^{\bar{i}} &= x^i, \quad i = 1, 2, \end{aligned} \tag{3.19}$$

where a, b, c, d are arbitrary constants and the choice $a = b = c = 1$ and $d = -1$ reduces (3.19) to (2.1). We summarize this notation in Appendix A. Using matrix notation one can write (3.19) as

$$x^{\bar{\mu}} = T^{\bar{\mu}}{}_\nu x^\nu \quad \text{with} \quad T^{\bar{\mu}}{}_\nu = \frac{\partial x^{\bar{\mu}}}{\partial x^\nu}. \tag{3.20}$$

From demanding norm invariance of x under (3.19), (3.20) one concludes the covariant and contravariant metrics

$$g_{\bar{\mu}\bar{\nu}} = \begin{pmatrix} d^2 - c^2 & ac - bd & 0 & 0 \\ ac - bd & b^2 - a^2 & 0 & 0 \\ 0 & 0 & -1 & 0 \\ 0 & 0 & 0 & -1 \end{pmatrix}, \quad g^{\bar{\mu}\bar{\nu}} = \begin{pmatrix} a^2 - b^2 & ac - bd & 0 & 0 \\ ac - bd & c^2 - d^2 & 0 & 0 \\ 0 & 0 & -1 & 0 \\ 0 & 0 & 0 & -1 \end{pmatrix}. \tag{3.21}$$

Remember that the Heisenberg evolution equations for generic fields in Minkowski coordinates read

$$[P_\mu, \phi(x)] = -i\frac{\partial \phi}{\partial x^\mu}, \tag{3.22}$$

especially $P_0 = H$ generates evolution in time x^0. The Heisenberg equations translate easily to the general light cone coordinates by inserting $P_{\bar{\mu}} = T_{\bar{\mu}}{}^\nu P_\nu$ into (3.22). Note that evolution in $x^{\bar{0}} = T^{\bar{0}}{}_\nu x^\nu$ is generated by $P_{\bar{0}} = T^\nu{}_{\bar{0}} P_\nu$ just like P^- generates dynamics in x^+. Several choices of parameters in (3.19) are discussed in [8] and correspond to systems in GLC quantization moving relatively to the heat bath. In the following we specialize (3.19) to the case discussed in [71, 70, 60], that is setting $a = 1, b = 1, c = 0, d = 1$. Explicitly, (3.19) reduces to

$$\begin{aligned} x^{\bar{0}} &= x^0 + x^3, \\ x^{\bar{i}} &= \boldsymbol{x}^i, \end{aligned} \tag{3.23}$$

which are called oblique coordinates. Thus $x^{\bar{0}} = x^+$ is evolved by $P_{\bar{0}} = P_0 = P^0$, i.e. one has

$$\left[P^0_{\text{LC}}, \phi(x)\right] = -\frac{i}{2}\frac{\partial \phi(x)}{\partial x^+}\bigg|_{x^3, \boldsymbol{x}_\perp \text{const}}, \tag{3.24}$$

where we used P^0_{LC} to distinguish the light cone energy operator from the instant form Hamiltonian. One can visualize this foliation of space-time by translating the light cone plane $\Sigma : x^+ = 0$ in positive and negative x^0 direction. But the key point is that the fields are still initialized and quantized on the light cone plane and all operators are constructed out of LC fields, including the energy operator $P^0_{\text{LC}} = \frac{1}{2}(P^+ + P^-)$. The components of the energy-momentum vector in oblique coordinates are

$$\begin{aligned} p_{\bar{0}} &= p_0 = p^0, \\ p_{\bar{3}} &= -p_0 + p_3 = -p_- = -\frac{1}{2}p^+, \\ p_{\bar{i}} &= p_i. \end{aligned} \tag{3.25}$$

Using the representation (3.25) one finds the dispersion relation

$$p^2 = g^{\bar{\mu}\bar{\nu}} p_{\bar{\mu}} p_{\bar{\nu}} = 2p_{\bar{0}}p_{\bar{3}} + p_{\bar{3}}^2 - p_{\bar{i}}^2 = -p^0 p^+ - p_-^2 - p_{\bar{i}}^2 = m^2, \tag{3.26}$$

still linear in $p_{\bar{0}}$. Quantization at equal $x^{\bar{0}}$ [59] is analogous to standard LC quantization since the dynamical system in the GLC frame written in lower components is still of first-order in $x^{\bar{0}}$ and the methods of Section 2.2 apply. Several simplifications, like the pole structure of the thermal propagators, emphasized in [8, 58], due the LC quantization survive the transformation to the GLC frame. These technical advantages of the GLC formulation may find future application in thermal field theory, e.g. the hard thermal loop approximation as suggested by Weldon.

Obviously the parameter choice above leads to a non-vanishing $g_{\bar{0}\bar{0}}$ metric component and the medium four-velocity in the (medium) rest frame can be written as

$$u^{\mu} = (\frac{1}{\sqrt{g_{\bar{0}\bar{0}}}}, 0, 0, 0), \tag{3.27}$$

and $u^2 = 1$ holds. Inserting the LC wave functions the statistical operator takes on the form

$$\varrho = \sum_{h} \sum_{n,n'} \exp\left\{-\beta \left(\frac{P_h^+}{2} + \frac{M_h^2}{2P_h^+} - \mu N(h,n) \right)\right\} \phi_{h/n} \phi^*_{h/n'} |h,n\rangle\langle h,n'|, \tag{3.28}$$

where $|n\rangle, \phi_{n/h}$ are the Fock states, belonging to the expansion of a (hadronic) eigenstate with mass M_h, and the wave function, accordingly. The function $N(h,n)$ summarizes all other charges where we have the prototype example of net quark number $N(h,n) = N_q(h,n) - N_{\bar{q}}(h,n)$ in mind. In Eq. (3.28) the hadronic states differing only in P^+ momentum are distinguished as in DLCQ. The partition function is central for the determination of thermodynamical quantities and the trace of (3.28) is best computed in the eigenbasis of the mass operator. In detail, it reads

$$\mathcal{Z} = \mathrm{Tr}\ \hat{\varrho} = \sum_{h,n} \exp\left\{-\beta \left(\frac{P_h^+}{2} + \frac{M_h^2}{2P_h^+} \right) - \mu N(h,n) \right\}. \tag{3.29}$$

We will utilize (3.29) to compute the pressure of an ideal gas in the next section and for interacting systems in Section 4.4.

Let us review briefly the perturbative setup of general LF thermal field theory [8, 57]. We consider the perturbative expansions of the fully dressed thermal propagators since the generalization to N-point functions is straightforward. Three aspects will be highlighted: the real-time thermal field theory, spectral functions and the connection to imaginary time formalism. For this survey we confine ourselves to scalar particles, for the generalization to non-scalar particles we refer the reader to [57] and the references therein.

The GLC real-time propagator is defined by

$$\begin{aligned} D_{\mathrm{LC}}(x,x') &= \langle \mathrm{T}_{\bar{0}} \left(\Phi_{\mathrm{LC}}(x) \bar{\Phi}_{\mathrm{LC}}(x') \right) \rangle_T = \frac{1}{\mathcal{Z}} \mathrm{Tr} \left\{ e^{-\beta P^0_{\mathrm{LC}}}\ \mathrm{T}_{\bar{0}} \left(\Phi_{\mathrm{LC}}(x) \bar{\Phi}_{\mathrm{LC}}(x') \right) \right\} \\ &= \frac{1}{\mathcal{Z}}\ \mathrm{Tr} \left\{ e^{-\beta P^0_{\mathrm{LC},f}} S^{-1}_{\mathrm{LC}} \mathrm{T}_{\bar{0}} \left(\phi(x) \bar{\phi}(x') S_{\mathrm{LC}} \right) \right\} \end{aligned} \tag{3.30}$$

where the $\mathrm{T}_{\bar{0}}(\cdot)$ denotes the time-ordering with respect to $x^{\bar{0}}$ and $\mathcal{Z}$ is the partition function. In the second line of (3.30) the light cone interaction representation was introduced. There $P^0_{\mathrm{LC},f}$ is the free part of the LC energy and the light cone S-matrix (and its inverse) is given by

$$S_{\mathrm{LC}} = U_{\mathrm{LC}}(\infty, -\infty) = \mathrm{T}_{\bar{0}} \exp\left\{ -i \int_{-\infty}^{\infty} dx^+ \int d\tilde{x}\ \mathcal{P}^-_{\mathrm{LC},I} \right\}, \tag{3.31}$$

with the interaction part $\mathcal{P}^{-}_{\mathrm{LC},I}$ of the light cone Hamiltonian density. The interacting fields $\Phi(x)$ are replaced by the free fields $\phi(x)$ following

$$\Phi_{\mathrm{LC}}(x) = U_{\mathrm{LC}}(0,t)\phi(x)U_{\mathrm{LC}}(t,0). \tag{3.32}$$

A characteristic of thermal field theory is the non-canceling of the inverse S-matrix in (3.30). At zero temperature this term does not contribute because of the stability of the vacuum state, i.e.

$$S|0\rangle = \alpha|0\rangle \tag{3.33}$$

and α is a phase factor that can be ignored. This is not possible when the thermal average is taken and leads to the characteristic 'doubled' degrees of freedom in thermal field theory. This treatment is closely analogous to the instant form case with the difference that in the S-matrix definition the interaction part of P^0_{LC} is P^-_I in contrast to P^0_I. Because of the presence of the second S-matrix in Eq. (3.30) the propagators acquire 2×2 matrix structure and the different components stem from the various possible contractions in (3.30). Doing perturbation theory one has to know the vertices of the theory and the free propagators. One finds the free LC real-time propagators in momentum space (in closed time formalism) to be

$$\begin{aligned}
D^{++}_{\mathrm{LC}}(p) &= \frac{1}{p^2-m^2+i\varepsilon} - 2\pi i n_b(p_{\bar{0}})\delta(p^2-m^2), && (3.34)\\
D^{+-}_{\mathrm{LC}}(p) &= -2\pi i\left[\theta(-p_{\bar{0}})+n_b(p_{\bar{0}})\right]\delta(p^2-m^2), && (3.35)\\
D^{-+}_{\mathrm{LC}}(p) &= -2\pi i\left[\theta(p_{\bar{0}})+n_b(p_{\bar{0}})\right]\delta(p^2-m^2), && (3.36)\\
D^{--}_{\mathrm{LC}}(p) &= \frac{-1}{p^2-m^2-i\varepsilon} - 2\pi i n_b(p_{\bar{0}})\delta(p^2-m^2), && (3.37)
\end{aligned}$$

where $n_b(p_{\bar{0}}) = \left(\exp(\beta p_{\bar{0}})-1\right)^{-1}$ is the Bose-Einstein distribution. As usual only three components of the real-time propagator are independent because of the relation

$$D^{++}_{\mathrm{LC}} + D^{--}_{\mathrm{LC}} = D^{-+}_{\mathrm{LC}} + D^{+-}_{\mathrm{LC}}. \tag{3.38}$$

Alternatively, one can use the technically simpler imaginary time formalism for statistical systems in equilibrium. By rotation to Euclidean time $t \to i\tau$ the standard methods of zero temperature field theory can be utilized and only a one component propagator $G^{\beta}_{\mathrm{LC}}(\tau,\bar{x})$ needs to be considered. With the help of the cyclic property of the trace one derives the Kubo-Martin-Schwinger (KMS) condition for the bosonic propagator

$$G^{\beta}_{\mathrm{LC}}(\tau) = G^{\beta}_{\mathrm{LC}}(\tau+\beta). \tag{3.39}$$

Therefore the propagator is periodic which allows a decomposition into discrete Matsubara frequencies. The free bosonic imaginary time propagator is

$$G^{\beta}_{\mathrm{LC}}(\omega_n,\underline{p}) = \frac{1}{p_n{}^2-m^2} = \frac{1}{2\omega_n p_{\bar{3}} + p^2_{\bar{3}} - p^2_{\bar{i}} - m^2}, \tag{3.40}$$

with the Matsubara frequencies

$$\omega_n = 2\pi i T n,\ n \in \mathbb{N}. \tag{3.41}$$

Finally, the spectral function relates the imaginary time propagator, the analytic continuation thereof and real time propagator by the Lehman representation. The full spectral function exhibits important informations about the medium like thermal masses, widths and excitations of particles. The general definition of the spectral function is the thermal average of the commutator of interacting fields

$$A_{\rm LC}(x) = \frac{1}{\mathcal{Z}} \mathrm{Tr}\left\{ e^{-\beta P^0_{\rm LC}} \left[\Phi_{\rm LC}(x), \bar{\Phi}_{\rm LC}(0) \right] \right\}. \tag{3.42}$$

This expression becomes very simple in the free field theory case where the spectral functions in momentum space reads

$$A^f_{\rm LC}(p) = 2\pi\varepsilon(p_{\bar{0}})\delta(p^2 - m^2) = -\frac{\pi}{p_{\bar{3}}}\varepsilon(p_{\bar{3}})\delta(p_{\bar{0}} - p_{\bar{0},\rm on}). \tag{3.43}$$

The imaginary time propagator is restored by contour integration following

$$G^\beta_{\rm LC}(\omega_n, \underline{p}) = \int \frac{dp'_{\bar{0}}}{2\pi} \frac{A_{\rm LC}(p'_{\bar{0}}, \underline{p})}{\omega_n - p'_{\bar{0}} + i\varepsilon}, \tag{3.44}$$

and the analytic continuation of $G^\beta_{\rm LC}(\omega_n, \underline{p})$ is obtained by replacing ω_n by (the complex) $p_{\bar{0}} + i\varepsilon$. In a similar fashion the real-time propagator is

$$D_{\rm LC}(p) = i \int \frac{dp'_0}{2\pi} \frac{A_{\rm LC}(p'_0, \underline{p})}{p_{\bar{0}} - p'_0 + i\varepsilon} + n_B(p_{\bar{0}}) A_{\rm LC}(p_{\bar{0}}, \bar{p}). \tag{3.45}$$

Writing the commutator in (3.42) out one introduces the (Fourier-transformed) correlators $D^>_{\rm LC}(p)$ and $D^<_{\rm LC}(p)$. Furthermore one is able to establish the following relations

$$\begin{aligned} A_{\rm LC}(p) &= D^>_{\rm LC}(p) - D^<_{\rm LC}(p), \\ D^>_{\rm LC}(p) &= (1 + n_B(p_{\bar{0}}))\, A_{\rm LC}(p), \\ D^<_{\rm LC}(p) &= n_B(p_{\bar{0}}) A_{\rm LC}(p). \end{aligned} \tag{3.46}$$

Using these equations and inserting the partition of unity into light cone momentum eigenstates a practical representation of the spectral function is

$$A_{\rm LC}(p_{\bar{0}}, \underline{p}) = \frac{2\pi^4}{\mathcal{Z}} \left(1 - e^{-\beta p_{\bar{0}}}\right) \sum_{n,m} e^{-\beta E_n} \delta(\underline{p} - \underline{p}^m + \underline{p}^n) \langle n|\phi(0)|m\rangle\langle m|\bar{\phi}(0)|n\rangle, \tag{3.47}$$

where $P^0_{\rm LC}|n\rangle = E_n|n\rangle$. Note again that these equations hold for scalar particles but can be easily generalized to the non-scalar case.

3.3. The ideal relativistic quantum gases

In the Hamiltonian framework of LF quantization the non-perturbative computation of thermodynamical quantities is given by summations over exponentiated eigenvalues. The ideal gases of Fermi/Bose statistics are the perfect test cases of these methods. DLCQ leads to a matrix representation of the LC Hamiltonian which is finite dimensional since the K is restricted from above. With increasing K more and more accurate approximations to the full Hamiltonian are constructed. Nevertheless, errors coming from the finite quantization volume and the finite resolution have to be estimated, especially for thermodynamics. The feasibility of finite temperature calculations in DLCQ at finite K is tested in the ideal case and gives an impression of the systematic errors coming from the truncation by comparing to the text book result.

The free DLCQ Hamiltonian is

$$H = \sum_n \frac{m^2}{n} a_n^\dagger a_n, \tag{3.48}$$

where a_n satisfies a fermionic or bosonic algebra. The free Fock basis is readily identified with the partitions of a given K and because of the simple structure of (3.48) the generic eigenvalue is

$$e_p = \sum_i \frac{1}{p_i} \quad \text{if} \quad \sum_i p_i = K. \tag{3.49}$$

For fermions each part in a partition occurs only once while for bosons any part can occur arbitrarily often. The numerical results for the fermionic theory are depicted in Figure 3.1. One realizes that the exact value of the one-particle state converged at finite resolution but also the slow filling of gaps in the continuous part of the spectrum $M^2 \geq 4$. The largest eigenvalue e_{max} is

$$e_{\text{max}} = \sum_i^N \frac{1}{p_i} \quad \text{with} \quad p_1 \neq p_2 \neq p_3 \cdots \neq p_n. \tag{3.50}$$

The harmonic sum is asymptotically approximated by the logarithm $\ln N + \gamma_E$ where γ_E is the Euler-Mascheroni constant. This leads to the approximative value for M^2_{max};

$$M^2_{\text{max}} = K e_{\text{max}} \approx K \left(\ln \left(\sqrt{8K+1} - 1 \right) - \ln 2 + \gamma_E \right) / M_0^2. \tag{3.51}$$

The eigenvalues can be found without constructing the partitions explicitly. As explained in Appendix D.1 the partitions are obtained by a recursive algorithm that can be easily translated to the eigenvalues e_p or the partition function. Although in Fig. 3.1 the resolution is limited by $K = 50$ the following computations were made with all states up to $K = 110$ included. In the recursive approach we were able to perform computations with a maximal resolution of $K = 400$ in reasonable time. However, the recursive procedure cannot be generalized to the interacting case. We focus here on the direct computations assuming the Hamilton matrix is at hand, because that approach is relevant for the Schwinger model.

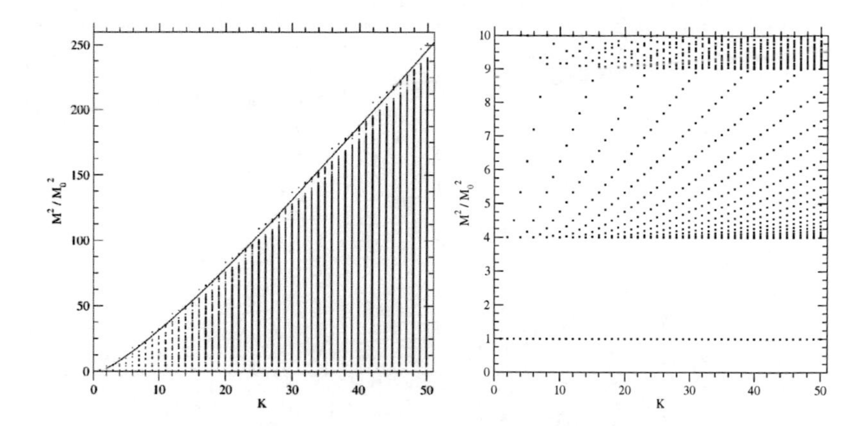

Figure 3.1.: (A) The normalized mass spectrum of a free fermionic theory given by the Hamiltonian (3.48). The solid line depicts an approximation to the largest eigenvalue. (B) A magnification of the lower part of the free spectrum $M^2 < 10M_0^2$ where $M_0 = m$ is the lowest mass. In particular the mass gap is shown.

We discuss now the numerical results of the thermodynamics of ideal gases. The textbook result for the thermodynamical potential (density) in the grand-canonical ensemble for vanishing chemical potential is

$$\omega_{f/b} = \mp T \int_0^\infty \frac{dp^+}{2\pi} \ln\left(1 \pm \exp\left\{-\beta\left(\frac{p^+}{2} + \frac{m^2}{2p^+}\right)\right\}\right), \tag{3.52}$$

where the upper/lower sign is for fermions/bosons (f/b). Equation (3.52) is derived analogously to the instant form case, except that the spatial volume gets replaced by the light-like extension L. Both expressions coincide for the potential densities. By summing the eigenvalues of the free LC Hamiltonian the exact analytical result (3.52) should be obtained for a range of parameters T, L. Because the potential Ω is extensive one expects a linear dependence

$$\Omega = -T \ln \mathcal{Z} = \alpha L + \beta \tag{3.53}$$

for large volumes. Figure 3.2 (A) shows the results for the free electron gas of mass $m = 0.5$ MeV at resolution $K = 110$. At small system volumes clear finite size effects are visible (see figure 3.2 (B)) and at large volumes there are derivations from the exact result (3.52) because of the finite resolution. These derivation due to the low resolution are reduced by an extrapolation to high values of K as explained in Appendix D.3. One has to identify a scaling window where the linear behavior (3.53) is present. Finding such a window is easy at small

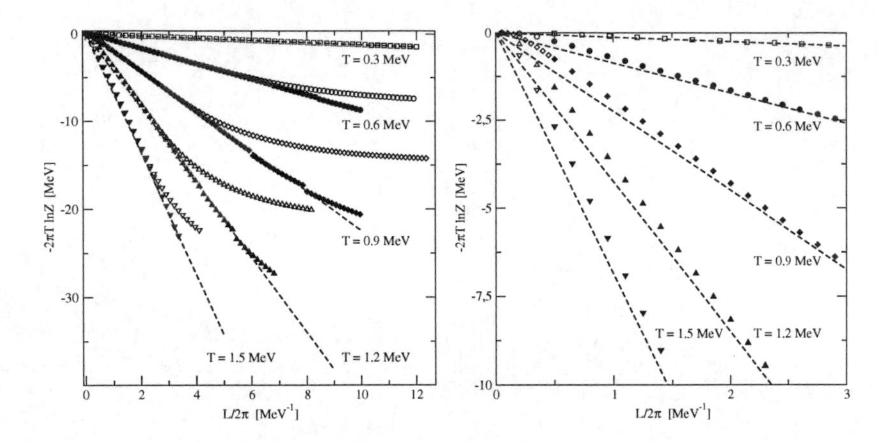

Figure 3.2.: (A) The thermodynamical potential $\Omega = -2\pi T \ln \mathcal{Z}$ as a function of the volume for different temperatures. Open symbols are the results for $\ln \mathcal{Z}$ at $K = 110$ while the closed symbols are extrapolated data points following Appendix D.3. Colored points were subject to a linear fit to the behavior in the bulk limit. The dashed line is the analytic result according to (3.52). The temperature and the box size is given in MeV because the mass of free electron is fixed to $m = 0.5$ MeV. (B) Zoom-in view of the small box length region showing the finite size effects. Same symbol coding as in Fig. A.

temperatures, but for increasing temperatures the scaling window is pushed to regions of small volumes. For the largest temperature shown in Figure 3.2 the relative error $\varepsilon_\Omega = (\Omega_f - \Omega)/\Omega_f$ is below 1.5%. Keeping this upper bound on the error we have not extended these calculations to higher temperatures, although possible. However, even if one takes the worst values of $\omega(T, L)$ at fairly small box sizes the relative error never exceeds 10%. The parameter intervals for T and L and the estimated error are taken as guiding lines for the interacting case in Section 4.4.

In the rest of the section we consider the possibility of defining an absolute light front temperature $T_{\rm LC}$ and other thermodynamical quantities in the micro-canonical ensemble. We are not trying to identify these quantities with the relativistic proper temperature, etc. Having the LC Hamiltonian matrix at hand one may try the standard recipe

1. compute density of states $\omega_{\rm LC}(p^-)$ from the light cone Hamiltonian for various values of the volume L

2. find the entropy $S(p^-, L) = \ln \omega_{\rm LC}(p^-)$

3. invert function $S(p^-, L)$ for $U(S,L) = P^-(S,L)$

4. compute thermodynamical quantities like absolute light cone temperature, pressure, heat capacity by

$$T_{\rm LC} = \left(\frac{\partial U}{\partial S}\right)|_L \,, \quad (3.54)$$

$$p = \left(-\frac{\partial U}{\partial L}\right)|_S \,, \quad (3.55)$$

$$c_L = \left(\frac{\partial U}{\partial T}\right)|_L \,. \quad (3.56)$$

A peculiarity arises observing the density of states $\omega_{\rm LC}(p^-)$. Inspecting the mass spectrum Figure 3.1 the function $\omega_{\rm LC}(p^-)$ diverges at $p^- = 0$ because of the infinite number of multi-particle states accumulating at $p^- = \frac{M^2}{p^+} = 0$. This leads to a divergent entropy for $U \to 0$. Proceeding to the light cone temperature one finds that $T_{\rm LC}$ is negative and vanishes in the limit $U \to 0$ and $S \to \infty$. Accordingly, this violates the third law of thermodynamics because the entropy is not vanishing nor constant at $T_{\rm LC} = 0$. Interestingly, also the (sub)additivity of entropy and internal energy of the union so-defined LC thermodynamical systems cannot hold at the same time. One notes that the setup above is valid for equilibrium states with respect to the light cone time, which are in general different from the instant form equilibrium states. Thus, there is no surprise that the LC thermodynamic functions differ from the standard ones. Obviously, the LC pendant of thermodynamics has unfamiliar properties distinct from relativistic thermodynamics and it is unclear to the author whether such a formalism can prove useful in some systems that are static in light front time. We leave this line of reasoning here and use the general light cone frame in the following applications.

4. Quantum Electrodynamics in 1+1 dimensions

This chapter contains the various obtained results at zero and finite temperatures in the LC quantized massive chiral Schwinger model. The number of publications on the Schwinger model is legion and all kinds of techniques in quantum field theory have been applied to it. The massless model is exactly solvable and shows many non-perturbative phenomena also known to exist in higher dimensional gauge theories like QCD. For the massive fermions one has to resort to perturbative or numerical approximations.

In the first part a collection of known facts about the massless and massive Schwinger model will be presented, without going in too much detail. These insights were found in different approaches like bosonization of two-dimensional fermion field theories or the lattice gauge theory and the interested reader is advised to the original literature [75, 76, 77]. It will hopefully become clear why the model has attracted much interest since Schwingers seminal publication [78]. The next section is concerned with the LF realization of the model and especially how the occurrence of theta vacua are compatible within the paradigm of a trivial LF vacuum state will be discussed. More subtle points on the LF massive Schwinger model are given, some of them are still subject of ongoing debate. For vanishing background field the DLCQ Hamiltonian is derived and results for the two lowest mass eigenstates for various couplings are extrapolated to the continuum limit. The structure functions of these bound states are computed and the picture of the Schwinger boson as a fermion-anti-fermion bound state is confirmed. Finally, the DLCQ matrices are used to compute the partition function and subsequently thermodynamical quantities. By carefully considering the thermodynamical as well as the continuum limit more sensible results than in Ref. [79] are found. In particular, the conjectured second order phase transition at $T/g \approx 1/\sqrt{\pi}$ is not confirmed.

4.1. QED$_{1+1}$ or the massive Schwinger model

The massive Schwinger model has many features known to be present also in QCD like quark confinement and chiral symmetry breaking. In contrast to QCD these properties are well-understood in the Schwinger model. Lessons learned in two-dimensional toy models can not directly be carried over to field theory in physical space-time since two dimensions offer some artificial simplifications, e.g. a non-dynamical gauge field. However, the solution of toy models in lower dimensions can provide tractable examples which can then be used as guidelines for the higher dimensional case.

Consider the action of quantum electrodynamics in two space-time dimensions

$$S = \int d^2x \mathcal{L} = \int d^2x \left(\bar{\psi} \not{D} \psi - m \bar{\psi} \psi - \frac{1}{4} F^{\mu\nu} F_{\mu\nu} \right), \tag{4.1}$$

where the covariant derivative is given by

$$D_\mu = i\partial_\mu + gA_\mu \tag{4.2}$$

and the components of the abelian field strength 2-form are

$$F^{\mu\nu} = \partial^\mu A^\nu - \partial^\nu A^\mu. \tag{4.3}$$

This model is super-renormalizable since the coupling has obviously mass dimension, it requires no infinite renormalization, and has finite bare parameters m and g. Beyond these parameters the massive Schwinger model depends on the angle variable θ. For an intuitive argument of how the θ-angle arises we follow [76, 80]. In axial gauge $A^1 = 0$ an θ-angle is found by examining the constraint given by the Gauss law

$$\partial_1^2 A^0 = g : \psi^\dagger \psi := j^0. \tag{4.4}$$

The Equation (4.4) is derived from $\delta S / \delta A^0 = 0$. The formal solution is

$$A^0 = \int dx^1 \, G \cdot j^0 - F x_1 - H, \tag{4.5}$$

where $G(x^1, x'^1) = \frac{1}{2}|x^1 - x'^1|$ is the Green function of (4.4) and F, H are arbitrary constants. By gauging away $A^1(x)$ and Eq. (4.5) the gauge field $A^\mu(x)$ has been fully eliminated and replaced by the linear Coulomb potential $G(x, x')$. The linear rising potential resembles the confinement of electrical charge in QED_{1+1}. The constant F is an electric background field.

Suppose such a background field to be present, e.g., created by a capacitor located at the ends of the one-dimensional volume. Then a ground state having fermion excitations is energetically favorable, since the electrostatic energy difference between the state containing a particle and antiparticle and the no-particle state is

$$\delta E = \frac{1}{2} \int dx^1 \, \left[(F \pm g)^2 - F^2 \right] = \frac{r}{2} \left[g^2 \pm 2Fg \right], \tag{4.6}$$

where r is the distance between the fermion and anti-fermion. Whether the upper or lower sign in Eq. (4.6) depends on the orientation of the particle pair. Thus, it is favorable to produce a fermion-anti-fermion pair if $|F| > \frac{g}{2}$ or expressed the other way around, one may apply a background field of strength $|F| \leq \frac{g}{2}$ without changing the ground state. Most importantly, if a field of strength F larger than $\frac{g}{2}$ is enforced pairs of fermions are produced and the field is reduced each time by g until $|F| \leq \frac{g}{2}$ holds. It is natural to define an angle to capture the periodicity of F, the θ-parameter

$$\theta = 2\pi F / g, \tag{4.7}$$

where $\theta \in [-\pi, \pi]$. The connection of the background field F to large gauge transformation will be exploited in the next section. At vanishing coupling g the ground state energy is given by the electrostatic energy and has a cusp at $\theta = \pi$. One concludes that a first order phase transition at $\theta = \pi$ occurs. In the massless limit, on the other hand, the vacuum energy is independent of θ. Therefore there exists a critical second order endpoint of the first order transition line. A numerical estimate [80] of the critical coupling is

$$(m/g)_c = 0.3335(2) \tag{4.8}$$

and the critical point has, within numerical errors, the critical exponents of the two-dimensional Ising model, i.e. $\nu = 1$ and $\beta = 0.125$.

The Schwinger model is exactly solvable in the limit of massless fermions in the sense that all exact Greens functions and the operator solution of the equations of motion are known [81, 82]. This property becomes obvious noting that the Schwinger model can be equally described by a free field theory of massive bosons. The correspondence between the fermion and boson formulation in two-dimensional theories will be further elaborated here. The charge-zero sector of the free massive Dirac theory in two dimensions can be equivalently written as sine-Gordon Hamiltonian [83]

$$\mathcal{H}_0 =: \left[\frac{1}{2}\Pi_\phi^2 + \frac{1}{2}(\partial_x\phi)^2 - Km^2\cos c\phi\right]:, \tag{4.9}$$

of a scalar field $\phi(x)$, supported by the definitions $c = 2\sqrt{\pi}$ and $K = \frac{1}{2\pi}e^{\gamma_E}$ with the Euler-Mascheroni constant γ_E. This extends to the interacting massive Schwinger model [76] and the Hamiltonian density of the bosonized massive Schwinger model becomes

$$\mathcal{H}_S =: \left[\frac{1}{2}\Pi_\phi^2 + \frac{1}{2}(\partial_x\phi)^2 + \frac{1}{2}\mu^2\phi^2 - Km\mu\cos(c\phi - \theta)\right]:, \tag{4.10}$$

where $\mu = \frac{g}{\sqrt{\pi}}$ is the Schwinger boson mass. For every local composition of fermionic operators an equivalent bosonic counterpart can be found. Important are the identifications

$$\begin{aligned}
i\bar{\psi}\partial\!\!\!/\psi &= \frac{1}{2}\partial^\mu\phi\partial_\mu\phi, & (4.11)\\
j^\mu &= \frac{1}{\sqrt{\pi}}\varepsilon^{\mu\nu}\partial_\nu\phi, & (4.12)\\
j_5^\mu &= \frac{1}{\sqrt{\pi}}\partial^\mu\phi, & (4.13)\\
\bar{\psi}\psi &= K : \cos c\phi :, & (4.14)\\
i\bar{\psi}\gamma_5\psi &= K : \sin c\phi :, & (4.15)
\end{aligned}$$

where $\varepsilon^{\mu\nu}$ is the Levi-Civita symbol in two dimensions. The scalar field may be interpreted with the help of (4.12) as the electric displacement field generated by the dynamical fermions and the vacuum polarization.

The massless Schwinger model has global chiral symmetry

$$\psi \longrightarrow e^{i\theta_0\gamma_5}\psi, \tag{4.16}$$

with $\gamma_5 = \gamma_0\gamma_1$. This symmetry gets explicitly broken by the bare fermion mass but utilizing (4.14) and (4.15) one recognizes that chiral transformations change the background field by the amount θ_0 [80]. Especially at $m = 0$ the fermionic and the bosonized Hamiltonian are chiral invariant and the ground state is infinitely degenerate in θ, thus the chiral symmetry is spontaneously broken. There is no massless Goldstone boson connected to chiral symmetry in the spectrum since the chiral current is affected by an anomaly. Allowing a finite bare fermion mass the "chiral current conservation" reads

$$\partial_\mu j_5^\mu = \partial_\mu \bar{\psi}\gamma_5\gamma^\mu\psi = -\frac{g}{2\pi}\varepsilon^{\mu\nu}F_{\mu\nu} + 2m\bar{\psi}i\gamma_5\psi. \tag{4.17}$$

One notes that the rhs of (4.17) is non-vanishing even in the limit $m \longrightarrow 0$.

The axial anomaly can be traced back to the fact that the theory, especially the current operator j_5^μ, is quantum mechanically not invariant under chiral transformations (4.16) and large gauge transformations. This may be rephrased in the path-integral language after introducing the θ-action

$$S_\theta = S - i\theta q(A), \quad \text{with} \quad q(A) = \int d^2x F_{12}, \tag{4.18}$$

together with the topological charge $q(A)$. Here F_{12} is the electric field in Euclidean space-time, i.e. with the substitution $x^0 \longrightarrow ix^2$. The anomalous breaking of chiral symmetry is caused by gauge field configurations carrying non-trivial topological charge [84]. Due to chiral symmetry breaking a fermion condensate $\langle\bar{\psi}\psi\rangle$ having non-trivial θ-dependence is formed and connected to the mass gap, i.e. the Schwinger boson mass μ in the massless limit is

$$|\langle\bar{\psi}\psi\rangle| = \frac{\mu}{2\pi}e^{\gamma_E}\cos\theta. \tag{4.19}$$

The last two prominent features of the Schwinger model to be discussed are the confinement of the fermions and the shielding of classical external currents. The shielding of external currents is established through the absence of a long-range force between two far separated charges of integer multiples of g. No shielding would occur for hypothetical fractional electrical charges [75]. The confinement of fundamental charge is realized through electric flux tubes formed between the fermions of opposite charges. When the fermion pair separates, the flux tube breaks and produces polarized fermion pairs in between the original pair. The new fermion pairs combine again to neutral bound states and the long range force of the original configuration gets screened.

For the background field $\theta = \pi$ the particles can move freely without experiencing any force if they are in alternating order, the fermion to left of the anti-fermion. This configuration is called half-asymptotic particle. In this case the field between neighboring particles is annihilated by the background field and it may be regarded as a signature of deconfinement in 1+1 dimensions [76]. The bosonic realization of half-asymptotic particles are solitons, namely kinks and anti-kinks in this theory. Following [76] these solutions of the field equations can be semi-classically (not taking loop corrections into account) motivated by looking at constant minima of the quantum potential of (4.10) at $\theta = \pi$

$$U(\phi) = \frac{\mu^2}{2}\phi^2 + K\mu m \cos(c\phi). \tag{4.20}$$

At small m/g a unique solution at $\phi = 0$ exist but at larger mass one recovers two degenerate minima at $\phi = \pm\frac{\sqrt{\pi}}{2}$. For other values of θ there always exists an unique minima at arbitrary coupling. One concludes that the Z_2 symmetry $\phi \leftrightarrow -\phi$ of (4.10) is spontaneously broken at weak coupling. The solitonic field configuration connecting these vacuum states monotonically is identified as the kink respectively the anti-kink. If well-separated kinks and anti-kinks are present they have to be in alternating order to ensure a global continuous field configuration. The $\phi \leftrightarrow -\phi$ symmetry is the charge conjugation symmetry in the fermionic description.

Topological gauge excitations and θ-vacua are also present in other field theories most notably QCD in three spatial dimensions. In general, topological gauge field configurations are possible in a field theory with gauge group $\mathcal{G}$ in $d+1$ dimensions if the d-th homotopy group $\pi_d(\mathcal{G})$ is non-trivial. Observing $\mathrm{U}(1) \sim S^1$ and $\pi_1(S^1) \sim \mathbb{Z}$ leads to the winding number $\nu \in \mathbb{Z}$ labeling the large gauge transformations. Thus, QCD_{1+1} shows no topological properties since every SU(N) group with N$\geq$ 2 has a trivial first homotopy group,

$$\pi_1(\mathrm{SU(N)}) = 0 \;\text{ for } N \geq 2. \tag{4.21}$$

In three spatial dimensions however QCD shows similar properties to the Schwinger model because the relevant homotopy group $\pi_3(\mathrm{SU}(3))$ is $\mathbb{Z}$. In particular the question of a non-zero vacuum angle in QCD is related to the strong CP problem and the hypothetical axions [85].

4.2. Light front massive Schwinger model

Zero modes and theta vacua

The Schwinger model was extensively studied in LC quantization in the past. Many efforts were made to understand the aspects discussed in the previous section, like the theta vacua, anomalous chiral symmetry breaking, the fermion condensate and confinement, also within the front form framework. On the light front a proper treatment of the gauge zero mode and the constrained quantization seems to be essential to recover the properties listed above. Furthermore this model has been one of the first test grounds for DLCQ. These computations in the chiral version of the model provided very convincing results for the LC Fock space approach to quantum field theories. The massless limit on the light front is questionable and the full operator solution of the Schwinger model can be obtained if one uses two light front plane $x^{\pm} = 0$ as initial and quantization surfaces, see [86]. However, the massive theory is well-defined and the finite box regularization allows one to separate the zero modes from the normal modes of the fields. The bulk limit $L \to \infty$ is taken at the end of any computations and interestingly some quantities like the electric background field connected to the θ-vacua and the mass spectrum are independent of the compactification volume. Introducing light front coordinates and the spinor projections $\Psi = \Lambda_+\Psi + \Lambda_-\Psi = (\psi_+, \psi_-)^T$ similar to (2.31) the Lagrangian (4.1) becomes

$$\begin{aligned} \mathcal{L}_{\text{LF}} =& i\psi_+^\dagger \overleftrightarrow{\partial_+} \psi_+ + i\psi_-^\dagger \overleftrightarrow{\partial_-} \psi_- + \frac{1}{2}\left(\partial_+ A^+ - \partial_- A^-\right)^2 \\ &- m\left(\psi_+^\dagger\psi_- + \psi_-^\dagger\psi_+\right) - \frac{g}{2}\left(j^+A^- + j^-A^+\right), \end{aligned} \tag{4.22}$$

where the normal-ordered current densities are $j^\pm = 2 : \psi_\pm^\dagger\psi_\pm :$ and, $A\overleftrightarrow{\partial_\mu}B = A(\partial_\mu B) - (\partial_\mu A)B$ holds. In the following we focus on the compact volume quantization of (4.22), i.e. we set $-L/2 \leq x^- \leq L/2$ and treat gauge field zero mode degrees of freedom as in Ref. [87].

By imposing periodic boundary conditions for the gauge fields and anti-periodic boundary conditions for the fermion fields,

$$\psi(-L/2) = -\psi(L/2), \tag{4.23}$$

$$A^\mu(-L/2) = A^\mu(L/2), \tag{4.24}$$

the separation of fields into zero modes and normal modes is done, see Section 2.4. Gauge fixing is maximally performed; we gauge away the normal mode $A_n^+ = 0$ and the zero mode $A_0^- = 0$. Note that strict light cone gauge $A^+ = 0$, discarding the gauge field zero mode A_0^+, is not allowed, since A_0^+ is gauge-invariant as explained in Section 2.4. The dependent degrees of freedom are the minus (so-called 'bad') components of the fermion field ψ_-, constrained by the non-dynamical Dirac equation

$$\left(2i\partial_- - gA_0^+\right)\psi_- = m\psi_+, \tag{4.25}$$

and the normal mode of the gauge field A_n^- constrained by the Gauss law

$$\partial_-^2 A_n^- = -\frac{g}{2}j^+. \tag{4.26}$$

Demanding the conditions (4.23) and (4.24), one determines the Greens functions of (4.25), (4.26) classically,

$$\begin{aligned} \mathcal{G}_a(x^- - y^-; A_0^+) &= \frac{1}{4i}\exp\left(\frac{-ig}{2}(x^- - y^-)A_0^+\right)\left[\varepsilon_L(x^- - y^-)\right. \\ &\quad \left. + i\tan\left(\frac{gL}{2}A_0^+\right)\right], \end{aligned} \tag{4.27}$$

$$\mathcal{G}_2(x^- - y^-) = \Xi_L(x^-y^-) = \frac{1}{2}|x^- - y^-| - \frac{(x^- - y^-)^2}{2L} - \frac{L}{6}, \tag{4.28}$$

where $\Xi_L(x^- - y^-)$ and other generalized functions (see [88]) are explained in Appendix B. Here $\mathcal{G}_a(x^- - y^-; A_0^+)$ is the Greens function of (4.25) and $\mathcal{G}_2(x^- - y^-)$ belongs to the Gauss law (4.26). Furthermore, the dynamical equations of motion read

$$2i\partial_+\psi_+ = m\psi_- + gA_n^-\psi_+, \tag{4.29}$$

for the ψ_+ field and for the zero mode A_0^+

$$\partial_+^2 A_0^+ + \partial_+\partial_- A_n^- = j^-. \tag{4.30}$$

After substitution of the dependent fields by the Greens functions one notes that the only remaining degrees of freedom are the zero mode A_0^+ and the fermion field ψ_+. Hence the zero mode presents a quantum-mechanical system coupled to the one-dimensional field theory.

The theory is canonically quantized at equal light cone time by demanding the anti-commutator

$$\left\{\psi_+^\dagger(x^-), \psi_+(y^-)\right\}_{x^+=y^+} = \frac{1}{2}\left(\delta_L(x^- - y^-) - \frac{1}{L}\right) = \frac{1}{2}\mathcal{D}_L(x^- - y^-), \tag{4.31}$$

where $\delta_L(x)$ is the finite volume analog of the Dirac delta distribution [88], cf. Appendix B. In the continuum limit $L \to \infty$ the $1/L$ finite volume correction on the rhs vanishes and one recovers the standard commutation relation. Utilizing the mode expansion of the Fermi field

$$\psi_+(x^-) = \frac{1}{\sqrt{L}}\sum_{n=1}^{\infty}\left(b_n e^{-ip_n^+ x^-} + d_n^\dagger e^{ip_n^+ x^-}\right) \tag{4.32}$$

leads to the basic anti-commutator algebra of the creation and destruction operators

$$\{b_n, b_m^\dagger\} = \{d_n, d_m^\dagger\} = \delta_{n,m}. \tag{4.33}$$

The bosonic gauge zero mode is quantized by requiring

$$\left[A_0^+, \Pi_{A_0^+}\right] = \frac{i}{L} \tag{4.34}$$

with the canonical momentum operator $\Pi_{A_0^+} = \partial_+ A_0^+$. The light cone momentum operators are constructed by integrals of the energy-momentum tensor which is defined by

$$T^{\mu\nu}(x) = \frac{\partial\mathcal{L}}{\partial(\partial_\mu\psi_r)}\partial^\nu\psi_r - g^{\mu\nu}\mathcal{L} \tag{4.35}$$

Specifically, for the Lagrangian (4.22) one computes

$$T^{++} = 2i\left(\psi_+^\dagger\partial_-\psi_+ - \partial_-\psi_+^\dagger\psi_+\right), \tag{4.36}$$

$$\begin{aligned} T^{+-} = {} & \Pi_{A_0^+}^2 - 2i\psi_-^\dagger\overleftrightarrow{\partial_-}\psi_- + 2m\left(\psi_-^\dagger\psi_+ + \psi_+^\dagger\psi_-\right) - (\partial_- A_n^-)^2 \\ & + gj^+A_n^- + gj^-A_0^+. \end{aligned} \tag{4.37}$$

These formulas, the Greens functions (4.27), (4.28) and the constraints (4.25), (4.26) enable one to derive the charge operator, the kinematical LC momentum operator and the LC Hamiltonian as

$$Q = \frac{1}{2}\int_{-L/2}^{L/2} dx^- j^+(x^-) = \int_{-L/2}^{L/2} dx^- : \psi_+^\dagger \psi_+(x^-) :, \tag{4.38}$$

$$P^+ = \frac{1}{2}\int_{-L/2}^{L/2} dx^- T^{++}(x) = 2i \int_{-L/2}^{L/2} dx^- \psi_+^\dagger(x^-)\partial_- \psi_+(x^-), \tag{4.39}$$

$$\begin{aligned} P^- &= \frac{1}{2}\int_{-L/2}^{L/2} dx^- T^{+-}(x), \\ &= \frac{L}{2}\Pi^2_{A_0^+} + \frac{m^2}{4}\int_{-L/2}^{L/2} dx^- dy^- \left[\psi_+^\dagger \mathcal{G}_a(x^- - y^-; A_0^+)\psi_+ + \text{h.c.}\right] \\ &+ \frac{g^2}{4}\int_{-L/2}^{L/2} dx^- dy^- \left(\psi_+^\dagger\psi_+\right)\mathcal{G}_2(x^- - y^-)\left(\psi_+^\dagger\psi_+\right). \end{aligned} \tag{4.40}$$

We return to the discussion of the local $U(1)$-gauge symmetry. The Hamiltonian is by construction invariant under local gauge transformations,

$$U = \exp\left(i\alpha(x)\right), \tag{4.41}$$

$$A_0^+ \rightarrow U A_0^+ U^\dagger - \frac{i}{g}(\partial^+ U)U^\dagger = A_0^+ + \frac{1}{g}\partial^+\alpha(x), \tag{4.42}$$

$$\psi_+ \rightarrow U\psi_+, \tag{4.43}$$

where $\alpha(x)$ is a arbitrary scalar function. These gauge transformations should be compatible with the conditions (4.23), (4.24). This is in particular true for α being periodic at the boundaries and the corresponding gauge transformations are called small, since these kind of transformations can be continuously deformed to unity. But there is a second class of gauge transformations obeying

$$\begin{aligned} \alpha(x^-) - \alpha(x^- - L) &= 2\pi\nu, \\ \partial^+\left(\alpha(x^-) - \alpha(x^- - L)\right) &= c, \end{aligned} \tag{4.44}$$

with ν a nonzero integer and c an arbitrary constant. The equations (4.44) are solved by α

being linear in x^- and the corresponding (large) transformations are

$$\begin{aligned} U_L &- \exp(i\frac{2\pi\nu}{L}x\ ), \\ A_0^+ &\rightarrow A_0^+ + \frac{2\pi}{gL}\nu, \\ \psi_+ &\rightarrow U_L\psi_+, \end{aligned} \tag{4.45}$$

where ν is the winding number of the transformation U_L. Implementing these special gauge transformations as unitary transformations in the quantum theory allows the definition of gauge-invariant (up to a phase factor) ground states which will turn out to be the theta vacua. In the zero mode sector the gauge transformations (4.45) act as constant displacement operators. Unitary operators realizing these displacements are

$$Z_\nu = \exp(i\nu\pi_0), \tag{4.46}$$

with the rescaled canonical momentum $\pi_0 = \frac{2\pi}{g}\Pi_{A_0^+}$. Let us also introduce the rescaled zero mode field $a_0^+ = \frac{gL}{2\pi}A_0^+$ and the combinations

$$z^\dagger = \frac{1}{\sqrt{2}}\left(a_0^+ - i\pi_0\right), \qquad z = \frac{1}{\sqrt{2}}\left(a_0^+ + i\pi_0\right), \tag{4.47}$$

which create ($z^\dagger$) and annihilate (z) zero mode boson quanta and obey the algebra $[z, z^\dagger] = 1$. On the vacuum state the operation (4.46) creates coherent states

$$|\nu, z\rangle = Z_\nu|0, z\rangle = \exp\left\{-\frac{\nu}{\sqrt{2}}\left(z^\dagger - z\right)\right\}|0, z\rangle, \tag{4.48}$$

where the vacuum state fulfills $z|0, z\rangle = 0$.

A similar construction can be done for the fermion sector with the important result that excitations with non-vanishing $+$-momentum in the LF vacuum can be present and still $\langle P^+\rangle = 0$ holds. Observing the commutator relation

$$\left[\psi(x), j^+(y)\right] = \psi_+(y^-)\delta(x^- - y^-), \tag{4.49}$$

one finds the following unitary operator

$$F_\nu = \exp\left\{i\frac{\pi}{L}\nu\int_{-L/2}^{L/2} dx^-\ x^- j^+(x^-)\right\}, \tag{4.50}$$

representing the large gauge transformation in the fermion sector. The operator allows again for the introduction of coherent states, analog to (4.48), as

$$|\nu, f\rangle = F_\nu|0, f\rangle = \exp\left\{\nu\sum_{m=1}^{\infty}\frac{(-1)^m}{m}\left(a_m^\dagger - a_m\right)\right\}|0, f\rangle, \tag{4.51}$$

where $a_m^\dagger, a_m$ are the Fourier coefficients of the current j^+, also called fusion operators. These can be given in terms of the elementary fermion creation and annihilation operators as

$$\begin{aligned} a_m &= \sum_{k=1}^{m} d_{m-k} b_k + \sum_{k=1}^{\infty} \left(b_k^\dagger b_{m+k} - d_k^\dagger d_{m+k} \right), \\ a_m^\dagger &= \sum_{k=1}^{m} b_k^\dagger d_{m-k}^\dagger + \sum_{k=1}^{\infty} \left(b_{m+k}^\dagger b_k - d_{m+k}^\dagger d_k \right), \end{aligned} \tag{4.52}$$

where $d_0 = b_0 = 0$ is understood. One can show that the coherent states $|\nu, f\rangle$ and $|\mu, f\rangle$ are normalized and orthogonal. The common ground state of the zero mode and the fermion sector is found as the product state $|\nu\rangle = T_\nu|0\rangle = (Z_1 F_1)^\nu |0\rangle = |\nu, z\rangle|\nu, f\rangle$. The vacuum state $|\nu\rangle$ is not invariant under T_μ and therefore the states

$$|\theta\rangle = \sum_{\nu=-\infty}^{\infty} \exp(-i\nu\theta) T_\nu |0\rangle \tag{4.53}$$

are constructed which pick up a phase factor under the transformations (4.45);

$$T_\mu|\theta\rangle = \sum_\nu \exp(-i\nu\theta) T_1^{\nu+\mu}|0\rangle = \exp(i\mu\theta)|\theta\rangle. \tag{4.54}$$

In particular, one proofs by direct computation the translation invariance of the theta vacua, i.e. $\langle\theta|P^+|\theta\rangle = 0$ and the average of the electric field operator

$$\frac{\langle\theta|\Pi_{A_0^+}|\theta\rangle}{\langle\theta|\theta\rangle} = \frac{e\theta}{2\pi} \tag{4.55}$$

to recover the standard interpretation of the theta vacuum state as strength of the background electric field. Thus, the front form pendant of the θ-vacua is found but the simple spatial translation invariance $P^+|\theta\rangle = 0$ is broken. Note that only the zero mode sector contributes to (4.55) and only fermion operators occur in P^+. Leaving away the fermionic excitation (4.51) in the θ-vacuum one establishes the translational invariance again. Whether contributions of the fermions constitute a difference to the instant form treatment is not clear so far.

It is noted for completeness that the theta vacuum has been subject of several investigations in the massive and massless LC Schwinger model. The influence of the gauge field zero mode in the coherent state representation was also investigated in the bosonic formulation of massless Schwinger model [89] and in the massive version [90] closely along the reasoning leading to (4.48) and (4.51). The use of coherent states to describe the quantum zero modes is not mandatory as the expositions in [91] in the bosonic massless Schwinger model show. There the coordinate representation of the free particle, i.e. eigenstates of A_0^+ were utilized to describe the zero mode dynamics. So-called N-vacua representing the filled Dirac sea were introduced in [92, 93] by

$$|0\rangle_N = \prod_{n=1}^{N-1} b_n^\dagger \prod_{i=1}^{\infty} d_i^\dagger |0\rangle. \tag{4.56}$$

These states undergo the transformations $|0\rangle_N \to |0\rangle_{N+1}$ when large gauge transformations (4.45) are applied. Likewise the zero mode is shifted by a constant. Normal-ordering is defined according to the states $|0\rangle_N$, that are again degenerate ground states. Implementing gauge invariance and charge neutrality lead to theta vacua and spontaneous breaking of chiral symmetry as usually. The same argument also works in the bosonized version of the massless Schwinger model [94] with the difference that the degenerate ground states are exclusively gauge field zero mode excitations.

Concerning the chiral condensate and the anomaly, the bosonized LF Schwinger model was used to derive the known relations (4.17), (4.19), see e.g. [90, 95]. The axial anomaly can be derived in the bosonized model using the equations of motion and, in the fermionic model, by defining a gauge-invariant current operator regularized by point splitting. In Ref. [92] the anomaly has been enforced by an additional classical constraint and the equations of motion. Yet one did not succeed to prove the same properties in the fermionic formulation of the massive Schwinger model using only the dynamical fermions. This is related to the inability to define a satisfactory chiral symmetry transformation for the two-component spinor fields that is compatible with the constraint (4.25). However, we refer to the discussion of LC chiral symmetry in [10]. Since, as emphasized in Section 4.1, the θ-vacua and chiral transformations are intimately related one should try to properly include zero modes of the Dirac operator $\partial_- - gA_0^+$. In the present case these zero modes are automatically excluded by the anti-periodic boundary conditions of the fermion fields and the Hamiltonian (4.40) is by construction chiral invariant.

DLCQ representation

For the practical DLCQ computations we have ignored the gauge field zero mode A_0^+, the invariance under large gauge transformations and the non-perturbative structure coming from the theta vacua as it is usually done in the literature [30, 96, 97]. This is regarded as forcing the background field to zero by hand. Interesting effects like the phase transition at $\theta = \pi$ and the θ-dependence of the bound state masses can not be investigated within this approach. The modified model is then called the LF chiral massive Schwinger model [86], since one chiral fermion field ψ_+ is singled out. Leaving aside the gauge field zero mode leads to changes in the LC Hamiltonian, mainly due to the simplification of the Greens functions (4.27) and (4.28). We set strict light cone gauge $A^+ = 0$ and the constraint equation (4.25) reads

$$2i\partial_-\psi_- = m\psi_+, \tag{4.57}$$

while in (4.26) the normal mode A_n^- turn into the full A^- field. Consequently the Greens function $\mathcal{G}_a$ in Eq. (4.27) is replaced by

$$\mathcal{G}_1(x^- - y^-) = \frac{1}{4i}\varepsilon_L(x^- - y^-). \tag{4.58}$$

Furthermore the Green function (4.28) of the Gauss law and the structure of (4.39) and (4.40) is kept. Inserting the mode expansion (4.32) one finds after quite a tedious computation

$$Q = \sum_n \left(b_n^\dagger b_n - d_n^\dagger d_n\right), \tag{4.59}$$

$$P^+ = \frac{2\pi}{L}\sum_n n\left(b_n^\dagger b_n + d_n^\dagger d_n\right), \tag{4.60}$$

$$P^- = \frac{L}{2\pi}H = \frac{L}{2\pi}\left\{\sum_n \frac{m^2}{n}\left(b_n^\dagger b_n + d_n^\dagger d_n\right) + \frac{g^2}{\pi}V\right\}. \tag{4.61}$$

The interaction term is divided into a normal-ordered part V_N and terms arising in the normal-ordering process, which are the induced inertia V_I. The normal-ordered part of the interaction is given by [30]

$$\begin{aligned} V_N = \sum_{k,l,m,n=1} \Big\{ & \left(b_k^\dagger b_l^\dagger b_m b_n + d_k^\dagger d_l^\dagger d_m d_n\right)[k-n|l-m]/2 \\ & + b_k^\dagger b_l d_m^\dagger d_n\Big([k+m|-l-n] - [k-l|m-n]\Big) \\ & + \left(b_k^\dagger d_l^\dagger d_m^\dagger d_n + d_n^\dagger d_m d_l b_k\right)[k+m|l-n] \\ & + \left(d_k^\dagger b_l^\dagger b_m^\dagger b_n + b_n^\dagger b_m b_l d_k\right)[k+m|l-n]\Big\}, \end{aligned} \tag{4.62}$$

where the brackets $[\,\cdot\,|\,\cdot\,]$ are defined as

$$[\,m\,|\,n\,] = \int_{-\pi}^{\pi} dx^- dy^- \; e^{imx^-}\mathcal{G}_2(x^- - y^-)e^{iny^-}, \tag{4.63}$$

leading to the following set of equations

$$\begin{aligned} [\,0\,|\,0\,] &= 0, \\ [\,0\,|\,m\,] &= [\,m\,|\,0\,] = 0, \\ [\,n\,|\,m\,] &= \frac{1}{n^2}\delta_{n+m,0}, \end{aligned} \tag{4.64}$$

see also the remarks on the symmetrization of the action of the Green function in the LC Hamiltonian in Ref. [96]. Note that we have not used the symmetrization prescription to arrive at (4.64) but instead inserted the Green function with removed zero mode contribution (4.28) into the definition (4.63). One recognizes the instantaneous photon propagator in the last line of (4.64). The induced inertia are computed as

$$V_I = \sum_{n=1} I_n\left(b_n^\dagger b_n + d_n^\dagger d_n\right), \tag{4.65}$$

with the constants I_n;

$$
\begin{aligned}
I_n &= \frac{1}{2}\sum_{m=1}^{K}\left([n-m|m-n]-[n+m|-n-m]\right) \\
&\stackrel{K\to\infty}{=} -\frac{1}{2n^2}+\sum_{m=1}^{n}\frac{1}{m^2}.
\end{aligned} \tag{4.66}
$$

In Appendix C the corresponding graphical representations for the interaction (4.62) and (4.65) are plotted.

The Schwinger model can be bosonized following the formulas (4.11) to (4.15). The transformation to bosonic variables can also be given at the level of creation and annihilation operators. For vanishing mass the Schwinger model is a free bosonic theory and we introduce the Fock expansion for the scalar field ϕ

$$
\phi(x) = \sum_n \frac{1}{\sqrt{n}}\left(a_n e^{ip_n^+ x^-} + a_n^\dagger e^{-ip_n^+ x^-}\right). \tag{4.67}
$$

The current-current coupling term in (4.40) is casted into a mass term for the scalar field. Consequently one achieves the following form of the LC Hamiltonian

$$
H_b = m^2\sum_{n=1}\frac{1}{n}\left(b_n^\dagger b_n + d_n^\dagger d_n\right) + \frac{g^2}{\sqrt{\pi}}\sum_{n=1} a_n^\dagger a_n, \tag{4.68}
$$

where the mass term of the fermions is kept. Introducing bosonic creation and annihilation operators proportional to the fusion operators in (4.52) as

$$
a_n^\dagger = \frac{1}{\sqrt{n}}\left(\sum_{m=1}^{n-1} b_m^\dagger d_{n-m}^\dagger + \sum_{m=1}^{\infty} b_{n+m}^\dagger b_m - \sum_{m=1}^{\infty} d_{n+m}^\dagger d_m\right), \tag{4.69}
$$

the Hamiltonians given in (4.61) and (4.68) can be related [30]. In the limit of vanishing fermion mass the bosonized Hamiltonian produces the expected spectrum, but is rather unstable against perturbations. The fermionic and the bosonic Hamiltonian (4.61) and (4.68) differ only in the normal-ordering prescription, i.e. the induced inertias. In contrast to (4.66) one obtains in the bosonic case

$$
I_n^b = \sum_{m=1}^{n}\frac{1}{m^2}. \tag{4.70}
$$

We are interested in the mass spectrum and the bound state wave functions, therefore we have to diagonalize the mass matrix $M^2 = P^+P^- = KH$ for increasing harmonic resolution K. Obviously the L dependence of M^2 cancels out. The algorithm to follow is

(i) create the Fock space at fixed harmonic resolution K obeying the condition $Q|phys\rangle = 0$, see Appendix D.1 for details

(ii) compute the LC Hamiltonian (4.61), respectively the mass matrix, see Appendix D.2

(iii) diagonalize M^2, or at larger resolutions compute the lowest eigenvalues and eigenvectors of M^2

(iv) compute observables of interest using the eigenvectors , e.g. structure functions of the relativistic bound states

The interaction strength is governed by the ratio m/g of the bare mass m and the coupling g. By convention the masses are measured in units of g in the lattice literature [80, 77, 98] while in the light cone literature [30, 97] a running interaction strength $\lambda^2 = 1/\left(1+\pi(m/g)^2\right)$ is introduced, casting the Hamiltonian in the form

$$H = \left(1-\lambda^2\right) H_0 + \lambda^2 V, \tag{4.71}$$

where the lowest mass is normalized to one. We will use both conventions, especially when comparing to the finite lattice results, but parameterize the coupling always by m/g.

4.3. Zero temperature

Spectra

Full diagonalization of the mass matrix is numerically feasible on work stations only for modest values of the harmonic resolution K. At higher resolutions only part of the spectrum can be determined by Lanczos or related algorithms, see Appendix D.2 for more details. A typical spectrum at coupling $m/g = 1$ is shown in Figure 4.1 (A). Although the whole picture in 4.1 (A) consists of discrete points one can easily distinguish the bound states or one-particle states of the interacting field theory , i.e. $M < 2M_0$ and the continuum part $M \geq 2M_0$ of the spectrum. The proliferation of the number of Fock states with increasing K is obvious but is mostly resulting in adding new scattering states to the spectrum. The bound state masses change marginally compared to the overall scale. Nevertheless the bound state masses are influenced by finite size effects as shown in Figure 4.1 (B) where the six lowest states are plotted solely and the data points are fitted by a quadratic polynomial in $1/K$. Only for the lowest mass eigenstate nearly no finite size corrections occur and M_0 is close to the continuum result for large K. For the other states the quadratic fit function is designed to take the first-order corrections due to the finite quantization volume into account. Utilizing the DLCQ approach one determines six bound states in the massive Schwinger model while the finite lattice Hamiltonian [80, 77] and the fast moving frame Hamiltonian [98] only quote results for the lowest two bound states. All these discrete states are excited states of the lowest mesonic state and can be further characterized by their behavior under parity transformation.

DLCQ can be used to obtain high-precision results about the physical spectrum, therefore the lowest two states are studied in more detail and the results are compared to masses computed by other means. Third, fourth and fifth order polynomial fits in $1/K$ are employed to

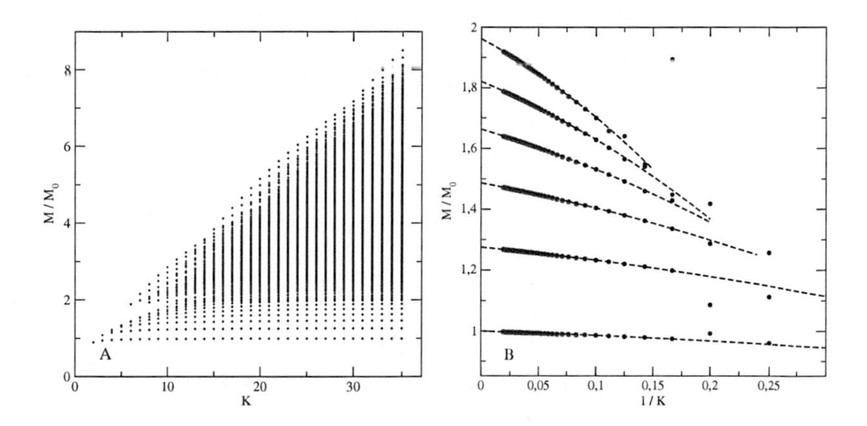

Figure 4.1.: (A) Mass spectrum computed for $m/g = 1$ at a maximum resolution of $K = 35$. The results for $K = 35$ consist of 7808 data points. (B) Six lowest mass eigenvalues and a quadratic fit to the data points colored in red for $m/g = 1$. Note the inverted scale on the x-axis with the continuum limit located at $1/K \to 0$. The lowest continuum mass is normalized to one.

extrapolate the chosen data points and estimate the error of the extrapolation heuristically. The different continuum values at $1/K = 0$ are averaged and the maximal deviation is used as an extrapolation error estimate. In Figure 4.2 the detailed study is depicted including the various fit functions. In panel (A) of Fig. 4.2 all fit functions seem to converge to the same continuum results and no deviations are visible. The right panels (B1) and (B2) show the mild variation of the different fits which lead to error estimates of $\delta M_0/g = 0.0007$ and $\delta M_1/g = 0.001$. The usual quantity to compare with other computations is not the mass but the binding energy E defined by

$$E/g = M/g - 2m/g. \tag{4.72}$$

The results are summarized in Table 4.1. The ground state mass is equal to three figures compared to most accurate results obtained by a finite lattice calculation in [80]. For the second mass state the presented results are one order of magnitude more accurate than the ones given in Ref. [77]. Comparison to the other methods listed in Table 4.1 is difficult since no numerical errors were given in [97, 30, 98]. These considerations are not confined to the coupling $m/g = 1$ and it is of course interesting to investigate how the spectrum changes from the weak-coupling fermionic theory to the strongly coupled bosonic theory. We follow the literature and tune the coupling in steps of powers of two from $m/g = 2^{-3}$ which is the small mass, strong coupling limit to $m/g = 2^5$, the small coupling, large mass limit. The behavior of the mass eigenvalues change drastically with the interaction strength. Only

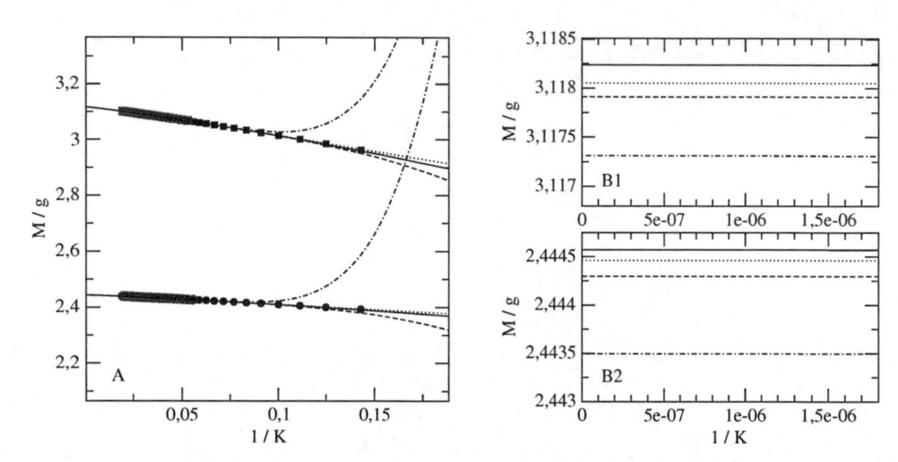

Figure 4.2.: (A) The lowest (circles) and second (squares) mass eigenstates for $m/g = 1$ as a function of $1/K$. Various polynomials in $1/K$ are used to fit the data points colored in red. The quadratic fit function is shown as solid line, the third order polynomial as dotted line, the fourth order as dashed and the fifth order as dash-dot line. (B1) Magnification of the limit region $1/K \approx 0$ for the second mass eigenstate. Variations of the continuum value for different fit polynomials is counted as error estimate of the extrapolation. (B2) Magnification of the limit region $1/K \approx 0$ for the first mass eigenstate.

the two couplings $m/g = 2^{-3}$ and $m/g = 2^5$ will be discussed here since they present the extreme situations met in these computation. The extrapolation properties of the masses in the intermediate coupling regime are similar to Fig. 4.2.

The results for the case close to the free fermion theory $m/g = 2^5$ are shown in Figure 4.3. One recognizes the strongly pronounced variation of the eigenvalues between odd and even resolutions K. This behavior comes from the different samplings of the strongly peaked valence wavefunctions for odd respectively even K. Only at high values of K a monotonic increase in $1/K$ is perceivable and subject to fits. But since only a handful of data points are available and these are located closely, the fit procedure is rather badly constrained. This fact causes the large errors of the extrapolation. It is interesting that the free fermion model, cf. Figure 3.1 (B) does not show this oscillating behavior and one observes that numerically small perturbations have strong impact on the eigenvalues.

We see the opposite situation for the nearly massless case at $m/g = 2^{-3}$. As shown in Figure 4.4 the masses at finite resolution do not oscillate and decrease when approaching the continuum limit. The masses converge very slowly in this interaction regime and the values for finite K are considerably distant from the masses computed by other methods. To

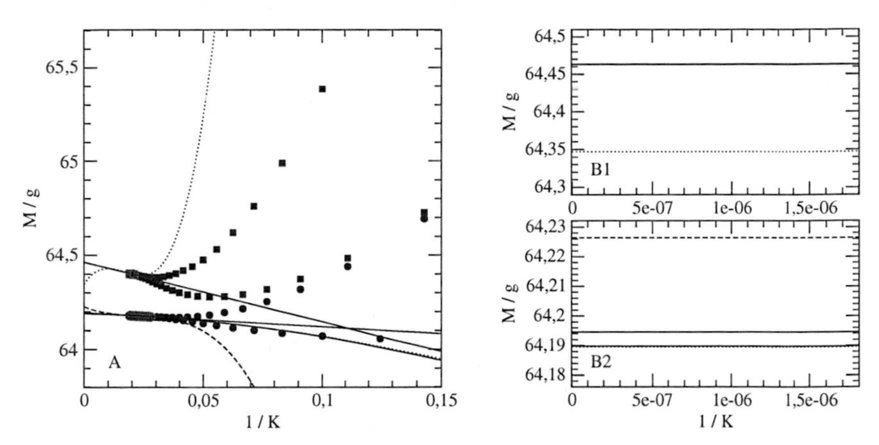

Figure 4.3.: (A) The two lowest mass eigenstates for $m/g = 2^5$ as a function of $1/K$. Same curve coding as in Figure 4.2. (B1) ((B2)) Magnification of the limit region $1/K \to 0$ for the second (first) mass eigenstate.

approach the correct values a sharp fall-off of the masses at small $1/K$ is necessary. This fall-off is indicated in Fig. 4.4 (A) and in particular in (B1), (B2) and (B3) one observes the large slope at $1/K \approx 0$. Moreover, the fall-off is not well described by polynomial fit functions used before. To this end the fit function is altered to contain besides the integer powers also fractional powers $\frac{1}{2}$, $\frac{3}{2}$ of $1/K$. Additionally, all data points have been used in the fit.

Looking for convergence improvement for the small mass regime we transformed (4.62) to boson variables. In practice the induced inertia term was changed to (4.70). The bosonized model is known to correctly reproduce the Schwinger boson mass. But including a small interaction term the model converges as slow as its fermionic counterpart. This is shown for the lowest mass state in Figure 4.4 (A) as triangle data points and in (B3), the close-up of the large K region. Again a rather large slope is present at the origin $1/K = 0$ and the extrapolation from the computed, finite K values is highly non-trivial.

From general considerations it is clear that the vanishing mass limit is problematic on the light front because physical particles are allowed to move inside the light front plane. In the massless Schwinger model an example of such degrees of freedom are the quanta associated with ψ_- field, often called left-movers. The dynamics of the left-movers is subtle and linked with taking the massless limit in the chiral model. See [86] for the suggestion to use initial conditions on two light front planes $x^\pm = 0$ to account for left-movers. In the massless case these additional initial conditions are necessary to obtain the correct operator solution of the Schwinger model known from the instant form. However, as it will be shown later for small but finite fermion bare masses the valence wavefunction is almost flat and the behavior at the endpoints $x = 0$ and $x = 1$ becomes important. One can improve the convergence of

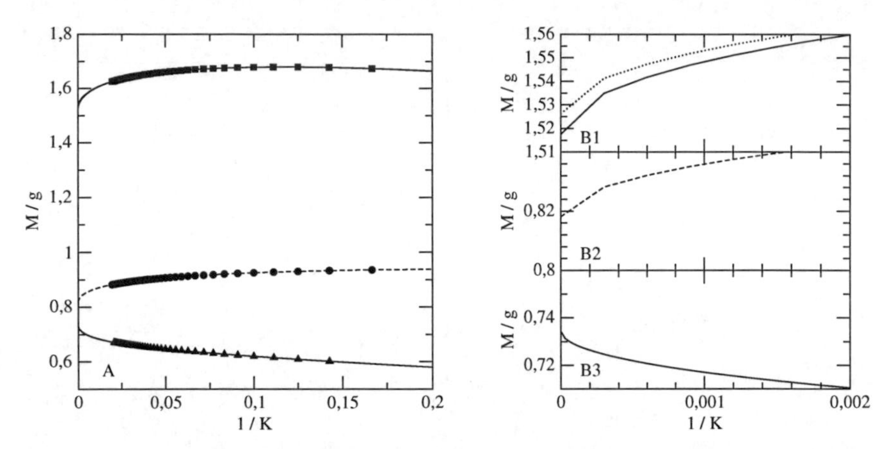

Figure 4.4.: (A) The first (circles) and second (squares) mass eigenstates for $m/g = 0.125$ as a function of $1/K$. In addition the lowest eigenstate of the bosonized Schwinger model (triangles) is included. All available data points are used in the fit. (B1) ((B2)/(B3)) Magnification of the limit region $1/K \to 0$ for the second (first) mass eigenstate of the (fermionic/ bosonic) Schwinger model.

the DLCQ computation for small masses by paying attention to this endpoint behavior of the wavefunction [99]. For the LC Hamiltonian (4.61) the terms containing the induced inertia I_n are supposed to be substituted by the following four-point interaction

$$\sum_{n=1} I_n \left(b_n^\dagger b_n + d_n^\dagger d_n\right) \longrightarrow \sum_{k,l=1} J_{kl}\, b_k^\dagger b_k d_l^\dagger d_l, \tag{4.73}$$

where the coefficients J_{kl} are given by the integral

$$J_{kl} = J_{lk} = \int_0^1 dy\, \frac{r^\beta(1-r)^\beta - y^\beta(1-y)^\beta}{(r-y)^2 r^\beta(1-r)^\beta} + \sum_{j=1}^{k+l-1} \frac{j^\beta(k+l-j)^\beta}{(k-j)^2(kl)^\beta} \tag{4.74}$$

with r denoting the ratio $k/(k+l)$ and β is some function of the coupling m/g. The results shown here were obtained in the unimproved DLCQ scheme and the convergence for different couplings was captured by trying different fit functions. Incorporating the improvement (4.74) is left for future work.

In summary, the results for all couplings are collected in Table 4.1 and confronted with results obtained by other methods like the finite lattice computations [77] and the density renormalization group computation for a finite lattice Hamiltonian [80], furthermore with the instant form Hamiltonian approach in a fast moving frame [98] and a LC variational computation [97].

Some comments on the different approaches are in order. The lattice Hamiltonian study in the Breit frame [98], moving with velocity $v = P/E$ close to the speed of light, used an instant-form quantized Hamiltonian in axial gauge $A^1 = 0$. It is discretized on a spatial lattice with up to $N = 384$ sites. This method is close to the DLCQ and similar approaches like [97] in the sense that a mode expansion of the field is employed.

A Tamm-Dancoff approximation to less than four particles was applied by Mo and Perry in Ref. [97]. The authors derived coupled mass eigenvalue continuum integral equations which are converted into matrix equations by expanding the wave function into (a finite set of) orthogonal functions like Jacobi polynomials, exponential and trigonometric functions. These lead to sensible results for the lowest two mass states since these are dominated by two- and four-particle Fock states in the Schwinger model. Finally, in [80, 77] a lattice gauge theory Hamiltonian is derived via the transfer matrix using compact gauge links. The corresponding Kogut-Susskind Hamiltonian reads

$$\begin{aligned} H_{\text{Lat}} =& -\frac{i}{2a}\sum_{n=1}^{N}\left[\phi^\dagger(n)U(n,n+1)\phi(n+1) - \phi^\dagger(n+1)U^\dagger(n,n+1)\phi(n)\right] \\ &+ m\sum_{n=1}^{N}(-1)^n\phi^\dagger(n)\phi(n) + \frac{g^2 a}{2}\sum_{n=1}^{N}L^2(n) \end{aligned} \tag{4.75}$$

where a is the lattice spacing and $U(n, n+1) = \exp(i\theta(n)) = \exp(-iagA^1(n))$ are the (compact, i.e. $\theta(n) = [0, 2\pi]$) lattice gauge links connecting sites n and $n+1$. The fermions are placed on a staggered lattice, meaning that for n even (odd) the $\phi(n)/\sqrt{a}$ converges to the upper (lower) component of the continuum fermion field $\psi_{\text{up}}(x)$ ($\psi_{\text{low}}(x)$). The variable $L(n)$ is conjugate to the lattice angular variable $\theta(n)$ and reduces to the electric field $E(x)$ in the limit $a \to 0$. In particular, in Ref. [80] the density matrix renormalization group is used to compute precise results for the lowest state at much larger lattices than the sizes that have been treated before, albeit the presence of the long-ranged Coulomb interaction. In the fifth row of Table 4.1 the results from lattice gauge theory computation are listed, the vector state mass is taken from [80] and the scalar state mass from [77].

Wave- and Structure Functions

The DLCQ algorithm not just enables one to compute the physical spectrum as the eigenvalues of the LC Hamiltonian but is in principle capable to deliver the full solution of the quantum field theory under consideration. The diagonalization of the LC Hamiltonian provides a spectral representation of the unitary time evolution operator or transfer matrix

$$U(x^+) = \sum_n e^{iE_n x^+}|\Psi_n\rangle\langle\Psi_n|, \tag{4.76}$$

m/g	this work	Mo and Perry [97]	Eller et al. [30]	Sriganesh et al. [77], Byrnes et al. [80]	Kroger and Scheu [98]
		ground state / vector state			
2^5	0.191(3)	0.201	0.201	0.194(5)	0.191
2^4	0.2366(8)	0.224	0.228	0.238(5)	0.235
2^3	0.2856(4)	0.288	0.280	0.287(8)	0.285
2^2	0.33933(5)	0.337	0.338	0.340(1)	0.339
2^1	0.39355(4)	0.393	0.393	0.398(1)	0.394
2^0	0.4442(7)	0.444	0.444	0.4444(1)	0.445
2^{-1}	0.4873(1)	0.488	0.488	0.48747(2)	0.489
2^{-2}	0.519(1)	0.520	0.534	0.51918(5)	0.511
2^{-3}	0.538	0.540	0.603	0.53950(7)	0.528
		first excited state / scalar state			
2^5	0.46(7)	0.458	0.458	0.45(1)	0.447
2^4	0.5623(3)	0.564	0.548	0.56(1)	0.559
2^3	0.696(4)	0.697	0.689	0.68(1)	0.690
2^2	0.839(2)	0.838	0.839	0.85(2)	0.837
2^1	0.9892(1)	0.989	0.985	1.00(2)	0.991
2^0	1.117(1)	1.119	1.126	1.12(3)	1.128
2^{-1}	1.2002(2)	1.201	1.228	1.20(3)	1.227
2^{-2}	1.21(3)	1.230	1.312	1.24(3)	1.279
2^{-3}	1.27(1)	1.219	1.407	1.22(2)	1.314

Table 4.1.: The eigenmasses of the ground state and the first excitation for different couplings m/g. For comparison the masses obtained by other methods are listed. The error given in braces is the variation of the last figure. See the text for further explanation.

where E_n is the eigenvalue of the n-th normalized eigenstate Ψ_n. The wavefunction is given as a decomposition in free Fock states as

$$|\Psi\rangle = \psi_{f\bar{f}}|f\bar{f}\rangle + \psi_{2f2\bar{f}}|2f2\bar{f}\rangle + \psi_{3f3\bar{f}}|3f3\bar{f}\rangle + \ldots, \tag{4.77}$$

where the ellipsis stands for the infinite number of higher Fock sectors. In (4.77) we have only included Fock states with charge zero because of the confinement property of the Schwinger model. With the transfer matrix available one may compute the correlation functions of the theory. For the Schwinger model the correlation function

$$\mathcal{C}(x_1, x_2, x_3, x_4) = \langle 0|\bar{\psi}(x_1)\bar{\psi}(x_2)\psi(x_3)\psi(x_4)|0\rangle \tag{4.78}$$

is the first non-trivial example describing (modulo the time-ordering) the propagation of a fermion-antifermion pair. With the help of the DLCQ solution, Fourier transformation and insertion of the time evolution operators the correlator is given by

$$\begin{aligned} &C_{n_1,n_2,n_3,n_4}(x_1^+,x_2^+,x_3^+,x_4^+) = \\ &\qquad \langle 0|U_1^\dagger(b_{n_1}+d_{n_1}^\dagger)U_{12}(b_{n_2}+d_{n_2}^\dagger)U_{23}(b_{n_3}^\dagger+d_{n_3})U_{34}(b_{n_4}^\dagger+d_{n_4})U_4|0\rangle \end{aligned} \tag{4.79}$$

with the shorthand notation $U_{ij} = U_i^\dagger U_j$ and $U_i = U(x_i^+)$. The integers n_1, n_2, n_3, n_4 denote momentum modes. Since the Schwinger model contains only three-particle vertices one can effectively condition the Fock states occurring in (4.79) as intermediate states. However, keeping track of all states generated during the time-evolution is a delicate computational task. Via the DLCQ approximation to the transfer matrix $U(x^+)$ the correlation functions will be explicitly volume dependent. This is also natural since any correlation is bounded by the box length L. Therefore one has to carefully control the convergence of the correlation function in the bulk limit in a similar way as it will be presented for the thermodynamical observables later. We did not compute the light cone correlation function being a side aspect of this work, but doing so would be an extension of the static bound state observables.

With the help of LC wavefunctions hadron observables can be calculated. The most important example is the momentum structure function of the bound state

$$G_n = \langle \Psi(K)|b_n^\dagger b_n|\Psi(K)\rangle, \tag{4.80}$$

$$\bar{G}_n = \langle \Psi(K)|d_n^\dagger d_n|\Psi(K)\rangle, \tag{4.81}$$

for particles and anti-particles. When properly normalized by $\int_0^1 dxG(x) = 1$, $G(x)$ describes the probability to find a fermion with momentum fraction x in the bound state. The momentum structure functions obey the sum rule

$$\sum_n (nG_n + n\bar{G}_n) = K. \tag{4.82}$$

In addition for a LC Hamiltonian symmetric under charge conjugation (as (4.61) is) one derives a second sum rule

$$\sum_n (G_n - \bar{G}_n) = 0. \tag{4.83}$$

In four dimensions the structure functions $G_n, \bar{G}_n$ are related to the parton model structure functions $q(x)$ observing that the Bjorken variable is $x_n = n/K$ and the structure function $q(x) = xG(x)$. These parton structure functions are inherently non-perturbative objects and hard to compute in a strongly coupled theory like QCD but directly measurable in deep inelastic scattering experiments. The calculation of the structure functions from first principles is one of the main motivations to formulate QCD on the light cone.

As a first step our presentation includes only the structure function and the valence wavefunction of the lower bound states. It is instructive to plot the valence wavefunction $\psi(x)$ which is the set of coefficients of the two-particle part of the LC wavefunction, i.e.

$$|\Psi_V\rangle = \int dx\, \psi(x)\, d_{1-x}^\dagger b_x^\dagger|0\rangle. \tag{4.84}$$

Especially the parity of the bound state can be made transparent looking at the valence wavefunction. As mentioned at the beginning the light cone framework is not a parity invariant setup since x^+ and x^- become interchanged under parity. The same holds for P^+ and P^-. Obviously the mass operator $M^2 = P^+P^-$ is invariant under parity which allows us to utilize parity as a quantum number of the states. In the center of mass frame the momentum fractions $x_i = k_i^+/P^+$ change into $y_i = k_i^-/P^+$ under parity using that P^+ is invariant. In the general N particle case this leads to

$$y_i = \frac{1}{x_i} \frac{1}{\sum_j^N \frac{1}{x_j}}. \tag{4.85}$$

For two particles one finds $y = 1-x$ so x and $1-x$ get interchanged by parity. Applied to the state in (4.84) one has a pseudo scalar when the valence wavefunction $\psi(x)$ is even or a scalar when $\psi(x)$ is odd under parity. Sometimes the pseudo scalar state is named vector state, as in Table 4.1, since there are no rotations in two dimensions.

All numerical results are obtained at a harmonic resolution $K = 51$ which leads to 50 data points on a equally distanced grid $x_n = n/K$. For strongly peaked or very fluctuating graphs we applied cubic splines to connect the data points.

The light cone valence wavefunctions are plotted in Figure 4.5 for the three couplings $m/g = 2^5, 1, 2^{-3}$. In all figures one recognizes that the wavefunction belonging to the lowest eigenstate is even and the second one is odd under parity. Hence the first state is pseudo scalar and the second one scalar. This classification extends to higher states for $m/g = 2^5$ and $m/g = 1$ only, as scalar and pseudo scalar states alternate. For the strongly coupled system the plotted higher states are all scalar. Since the wavefunctions are components of the DLCQ eigenvector of unit norm no specific normalization is applied to the valence sector alone.

In Figure 4.5 (B) the wavefunctions are centered around the point $x = 0.5$ because the particle and anti-particle share the total momentum equal parts. The free fermion eigenstates of the kinetic term of the Hamiltonian are of course sharp momentum states peaked at a certain x and vanish everywhere else. In comparison the wavefunctions in the Figure 4.5 (A) and (C) spread over the whole interval. In particular in Fig. 4.5 (A) the wavefunction vanishes at the endpoints without artificially enforcing this by hand. In contrast in Fig. 4.5 (C) this does not happen because the coarse grid of data points does not capture the fall-off at the endpoints. Observing this lead to the improvement suggestion (4.74). The third to sixth state at strong coupling consists mostly of two Schwinger bosons in relative motion. These have only small two-particle components, we will return to this point looking at the structure function of these states.

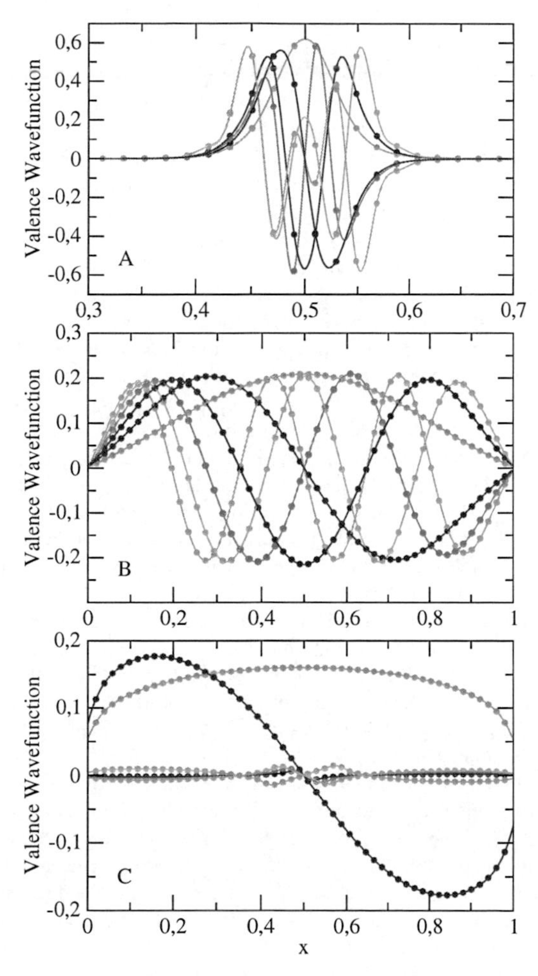

Figure 4.5.: The valence component of the light cone wavefunction of the lowest six states for the couplings $m/g = 2^5$ (A), $m/g = 1$ (B) and $m/g = 2^{-3}$ (C) for harmonic resolution $K = 51$ (increasing coupling or decreasing mass from top to bottom). The color coding is: first state brown, second state red, third state blue, fourth state green, fifth state orange, and sixth state turquoise. The dots are computed data points and the connecting curves are cubic splines.

To discuss the properties of the bound states we now turn to the structure functions. First we investigate how large the influence of the higher Fock components for the different couplings is. There is a large number of curves possible to be shown here (six structure functions for nine couplings each having six contributing Fock sectors). We limit the discussion to the structure function of the two lowest states for three couplings in Figure 4.6 and a detailed view on the higher Fock components thereof. One has contributions from Fock sectors containing up to 12 particles and anti-particles at a resolution of $K = 51$. For nearly all states shown the higher Fock sectors are strongly suppressed. Therefore we only include the four and six particle contributions and do not show higher ones. The graphs are normalized such that $\int_0^1 dx G_V(x) = 1$, where $G_V(x)$ is the valence structure function. We found for every coupling and all states that Fock states with particle content larger than six are suppressed roughly by a factor $1/1000$ (this translates into a suppression factor of 10^{-6} in the structure function). These findings are in accordance with the result of the variational LF Tamm-Dancoff approach for the ground state in [100], who optimized the wave function in the two-particle sector to contribute more than 99% to the total wave function for couplings $m/g < 1/2$. The only state where the four particle contribution is not totally negligible is the second state at strong coupling, Fig. 4.6 (C2). Although the results seem to justify a particle number cut-off *a posteriori* this would be an uncontrolled approximation to the full spectrum. In general, the eigenvalues and eigenvectors of LC Hamiltonian matrices are unstable against variations of single matrix elements and cuts. Nevertheless, the results shown in Figure 4.6 are in accordance with the paradigm of light cone field theory, that the bound state spectrum can be accurately described by the few-particle sectors of Fock space.

The broad distributions in Fig. 4.6 (A1), (A2) and (C1), (C2) have no simple interpretation in fermionic constituent particles but involve many plane wave fermion momentum states. The second states are the first spatial excitation of the formed bound state. If the structure function is sharply peaked, as in Figure 4.6 (B1), one concludes that the the bound state mainly consists of weakly interacting fermions. Obviously, Figure 4.6 (B2) is mainly a state where two free fermions are in small relative motion.

For further illustration two more figures are included, one is Fig. 4.7 (A) showing the dramatic change of the ground state structure function with increasing coupling. A similar plot in the valence approximation can be found in the first publication on the DLCQ Schwinger model [30]. Again one observes that increasing the interaction leads to a flatting out of the delta-like distribution of free fermions. In the massless limit the distribution function is constant on the whole interval, since the lowest state is one Schwinger boson having momentum $\frac{L}{2\pi} P^+ = K$ and by (4.69) every fermion momentum state gets equally occupied.

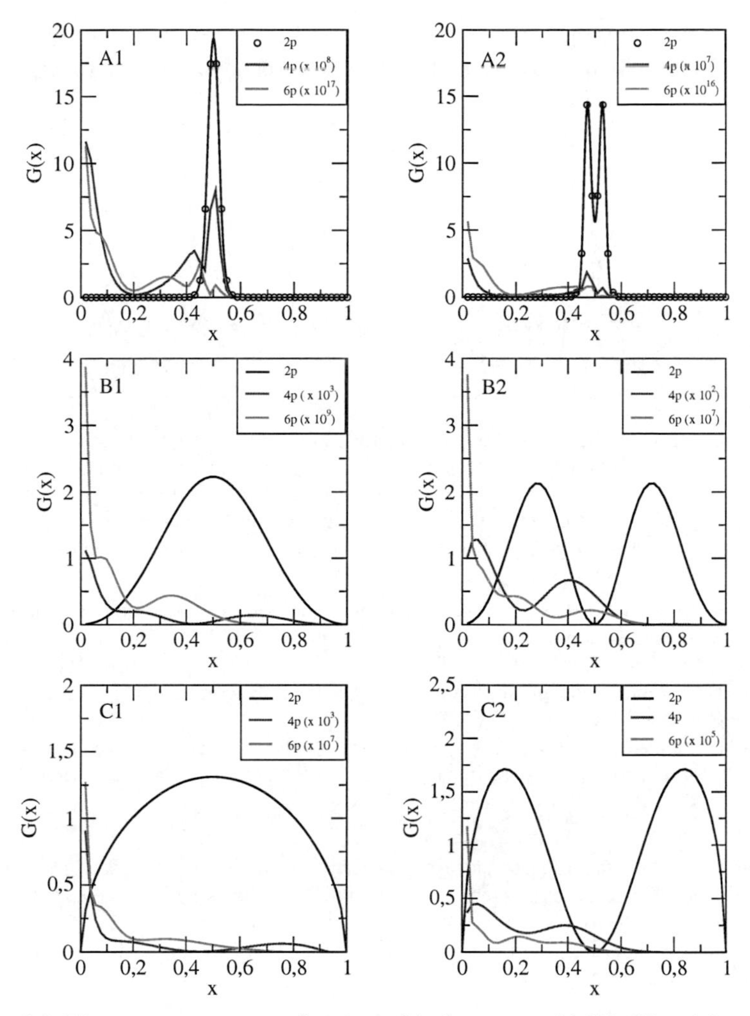

Figure 4.6.: The momentum structure functions of the lowest state A1 (B1, C1) and the next to lowest state A2 (B2, C2) for the interactions $m/g = 2^5$ $(1, 2^{-3})$. The Np denotes the contribution of the N-particle Fock sector. Note the enhancement factors for the different Fock sectors included in the legend box. In the panels A1, A2 a cubic spline connects the data points to guide the eye.

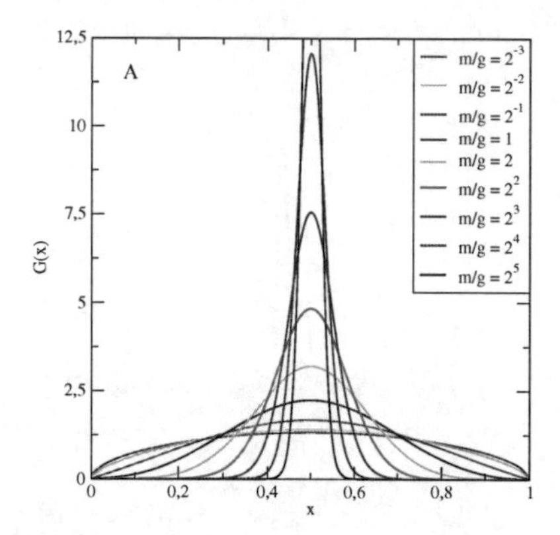

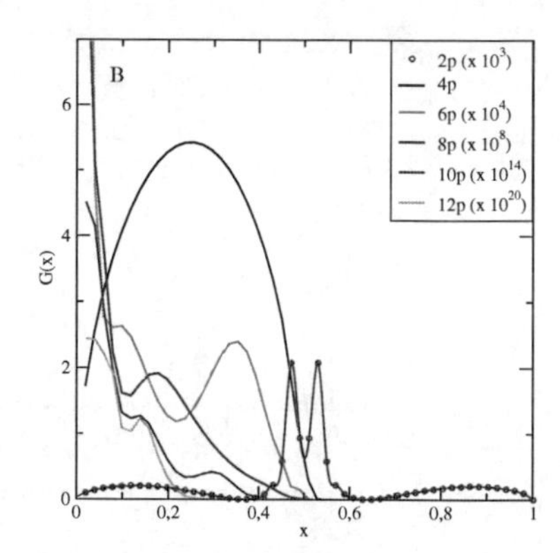

Figure 4.7.: (A) The momentum structure function of the lowest mass eigenstate. Interaction strength as indicated in the legend box. (B) The momentum structure function of the third mass eigenstate at strong coupling $m/g = 2^{-3}$. The major contribution comes from the four-particle sector and is centered around $x = 0.25$. The state contains two ground state mesons equally sharing the total momentum.

The Figure 4.7 (B) presents a state of the continuous spectrum for strong coupling. The four-particle contribution dominates the structure function while the other Fock sectors are suppressed at least by factors of 10^{-3}. The distribution is non-zero at the interval $0 < x < 0.5$ with a shape similar to the ground state Figure 4.6 (C1). We conclude that this state are two ground state mesons having each $K/2$ momentum.

Further interesting observables are hadron form factors. For example the elastic electromagnetic form factor can be given in terms of LC wave functions as the matrix element

$$\langle \Psi(P')| J^{\mu}_{\text{em}}(q) |\Psi(P)\rangle = F_{\text{em}}(-q^2)\,(P'+P)^{\mu}\,. \tag{4.86}$$

Evaluating the form factors on this level of elaboration would require to work with various LC wavefunctions for varying, preferably large, resolutions. This is not impossible but numerically challenging even in one dimension. The same challenge arises if one tries to calculate distribution amplitudes in the DLCQ scheme. The distribution amplitude for e.g. pseudo scalar mesons read

$$f_{\pi}\phi_{\pi}(x) = \int dx^{-}\,\langle 0|\psi^{\dagger}_{+}(0)\gamma_5\psi_{+}(x^{-})|\Psi(P)\rangle e^{ixP^{+}x^{-}} \tag{4.87}$$

where f_{π} is the decay constant. Although Eq. (4.86) and (4.87) also present important quantities most DLCQ computations restrain themselves to give results for the structure function and wavefunctions of the lower bound states since these are easier established, see

[30, 101, 97, 100] in the context of QED_{1+1} and QCD_{1+1}. For counterexamples to this rule see [12, 53].

4.4. Thermodynamical quantities

The computations at finite temperature follow the discussions in Section 3 and in particular the numerical results presented for non-interacting theories. The DLCQ Hamiltonian matrices are used to determine the partition function $\mathcal{Z}(T, L)$ of the canonical ensemble at different couplings, temperatures and volumes. The canonical ensemble is automatically implied in the Schwinger model by the charge neutrality condition of physical states. All states carrying charge have diverging LC energy and thus, zero statistical weight. All other thermodynamical quantities are derived from $\mathcal{Z}$ by differentiation with respect to the temperature T. Results are presented for the pressure

$$p = -\omega = \frac{T}{L} \ln \mathcal{Z} \tag{4.88}$$

and the internal energy

$$u = \frac{T^2}{L} \frac{\partial}{\partial T} \ln \mathcal{Z}. \tag{4.89}$$

Furthermore the entropy density is easily obtained by

$$s = \frac{1}{T} (u + p) , \tag{4.90}$$

as is the susceptibility

$$c_v = \frac{\partial}{\partial T} u. \tag{4.91}$$

In the DLCQ approach the partition function $\mathcal{Z}$ is computed directly contrary to lattice thermodynamics where the partition function, as the normalization of the path integral, is not calculated, but averages of gauge invariant operators are obtained. One consequence is that the pressure is not obtained unambiguously on the lattice. Equation (4.88) is discretized using the DLCQ matrices of the corresponding operators. Accordingly, for the partition function one obtains

$$\begin{aligned} \frac{1}{L} \ln \mathcal{Z} &= \frac{1}{L} \ln \left(\mathrm{Tr}\ \hat{\varrho} \right) \\ &= \frac{1}{L} \ln \left(\sum_{K} \sum_{p(K)} \exp \left\{ -\frac{\beta}{2} \left(\frac{2\pi}{L} K + \frac{L}{2\pi} \frac{M^2(K)}{K} \right) \right\} \right) , \end{aligned} \tag{4.92}$$

where T and L are given in units of the ground state mass M_0. The trace, running over the whole Fock space, is rewritten in Eq. (4.92) by utilizing momentum conservation, so $p(k)$ is the matrix dimension of the Fock sector of a specific K. We adopt the normalization by M_0

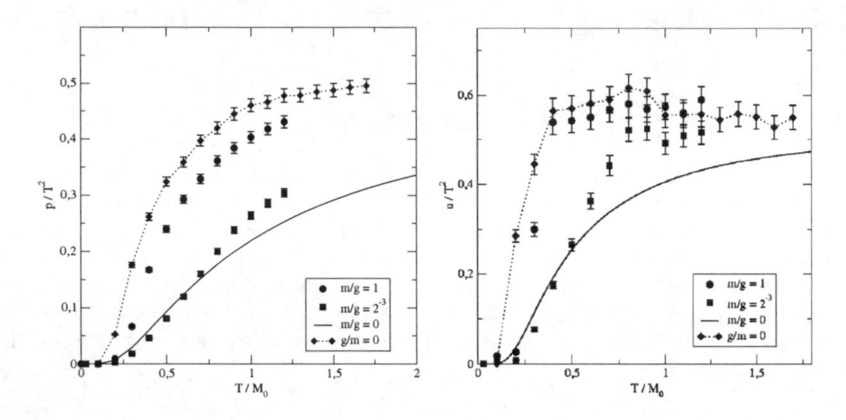

Figure 4.8.: (A) Dimensionless ratio of p/T^2 depending on temperature in units of the ground state mass M_0. (B) Dimensionless ratio of internal energy (density) and temperature depending on temperature T/M_0.

since the finite mass gap is the natural scale to be related to physical units by experiments. A maximal temperature of $T \leq 1.5M_0$ seems to be small but we like to stress the comparison to four-dimensional QCD where the according lowest state is the pion with a mass $M_\pi = 140$ MeV. Therefore the temperature regime handled here would be large enough to detect the confinement-deconfinement phase transition (at zero quark chemical potential) conjectured at $T_c \sim 190$ MeV.

From Section 4.2 we expect that the thermodynamical quantities interpolate between the two-component Fermi system for vanishing coupling and the Bose gas for vanishing fermion mass. The former expectation needs some modification, since the results for free gas and interacting gas are compared and the latter is confined to the total charge zero sector, we restrict also the free Fock states to the charge zero sector. This explains why the non-interacting Fermi gas was computed numerically instead of using the textbook result (3.52).

To evaluate (4.92) numerically the trace of the matrix exponential poses a hard computational problem. Exponentiation of matrices can be done in various ways [102] where the most straightforward one is the exponentiation of the eigenvalues. However, the computation of the eigenvalues is by itself costly for large matrices and delivers more information than is actually needed if one is interested in the trace only. If one refrains from the eigenvalues the matrix exponential has to be determined approximately. We follow the combination of two algorithms which are explained in detail in the Appendix D.3. The exponential $\exp(H_0 + V)$ is split by Trotter decomposition to first order

$$e^{H_0+V} = \left(e^{H_0/2n} e^{V/n} e^{H_0/2n}\right)^n + O(1/n), \tag{4.93}$$

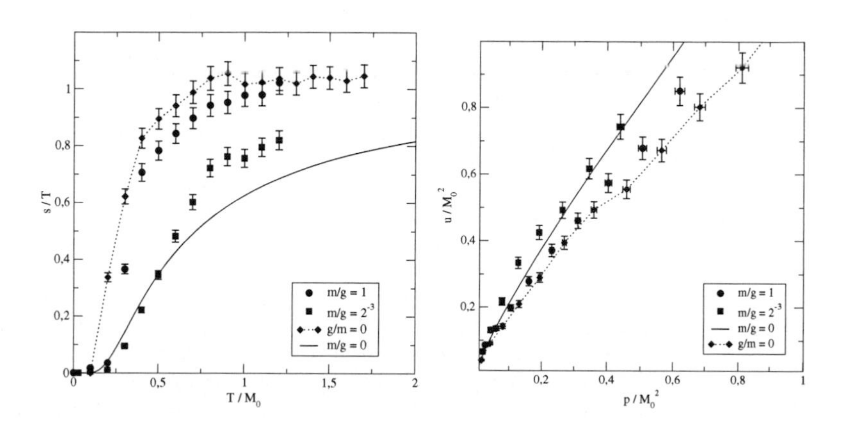

Figure 4.9.: (A) The dimensionless ratio s/T as a function of temperature T/M_0. (B) The equation of state, pressure vs. energy density, for various interaction strengths. Note the different normalization.

taking advantage of the fact that $\exp(H_0/2n)$ can be computed easily. The non-diagonal $\exp(V/n)$ is formally approximated by the Taylor series. Since the computation of matrix products is again not cheap in terms of computational time the trace is calculated stochastically. Applying the random vector method, the trace of an arbitrary matrix A is

$$\operatorname{Tr} A \approx \frac{1}{N} \sum_{n} \langle \xi_n | A | \xi_n \rangle, \tag{4.94}$$

where ξ_n is a random vector with components in $\{1, -1\}$. This method avoids matrix products and only matrix-vector operations are necessary. A drawback of this algorithm is that the parameter space (T, L) is not simply traversed. Like in the free case one is looking for the scaling window of linear dependence of ω on the box length L. Some pictures showing the L-dependence for different temperatures are given in Figure E.1 in Appendix E. In principle it is also useful to compute the thermodynamical potential at finite box size L as it is for finite systems with spatial volume in the instant form. However, the light-like system size L, instead of a space-like extension, is hard to interpret in terms of phenomenological thermodynamics. The equivalence of front- and instant-form thermodynamics is only recovered in the limit $L \to \infty$.

The first result is the ratio p/T^2 of the QED gas for different couplings as depicted in Figure 4.8 (A). Besides the curves for the couplings $m/g = 1$ and $m/g = 2^{-3}$ the idealized gases for $m/g = 0$ and $g/m = 0$ are included. One notes the monotonic change from the charge neutral free fermionic to the free bosonic system. Moreover, the idealized Bose gas and the case of strong interaction are located closely, in correspondence with the expectations

In the high temperature limit the free Bose gas pressure approaches $p/T^2 \to \pi/6$ which is derived by neglecting terms of the order m^2/T. This is equivalent to the ultra-relativistic limit. One notes that only the free fermion $g/m = 0$ in Fig. 4.8 (A) reaches the regime of the asymptotic value. For the other cases the numerical resolution is not yet enough to do precise computation at high temperatures. The situation is different for the internal energy depicted in Figure 4.8 (B). Again the high temperature limit is $u/T^2 \to \pi/6$ and one notes that the asymptotic value is fairly achieved by all curves. Therefore we conclude that the numerically accessible temperature interval is sufficient. The internal energy data points are obtained by a numerical derivative routine from the partition function. This explains the strongly fluctuating behavior. A more narrow temperature grid would cure this problem. Alternatively one could directly compute the statistical average of the light cone energy operator $\langle E_{LC}\rangle = \frac{1}{2}\langle P^+ + P^-\rangle$ but that would mean more numerical effort than simply differentiating.

Some more thermodynamical quantities are presented, namely the entropy density and the equation of state that can be compiled from the data points in the Figures 4.8 (A) and 4.8 (B). The entropy density determined by (4.90) is given in Fig. 4.9 (A) and shows qualitatively the features of the pressure plot including the distance of the graphs to the high temperature limit. The thermodynamical equation of state, pressure vs. energy density, is obtained by collecting pairs (p, u) at the same temperature. The plot of these pairs is given in Figure 4.9 (B). In comparison to the previous plots the pressure and energy is here given in units of the ground state mass (squared). All equations of state for the different couplings are located in a narrow band and offer the same qualitative behavior, so no large effect due to the interacting gas is seen. For intermediate pressure the equation of state can be written as $p \approx c(m/g)u$ where $c(m/g)$ is some constant depending on the interaction. We assume a conservative mean relative error of 2.5% for the pressure of interacting QED_{1+1} gas indicated in the Figure 4.8 (A) by error bars. For the energy density u we deduce a relative error of 5% by the approximative linear relation between p and u with $c = 2$. We restrain from giving results for the susceptibility since the graphs would be too noisy at the temperature grid presently used.

Some comments are added to what extent our results differ from findings in [79]. The authors of [79] used the DLCQ spectrum at harmonic resolution $K = 25$ to compute the partition function, indeed similar to the approach in (4.92). The crucial difference is that Elser and Kallionatis (EK) regard all (discrete) Fock states at a finite harmonic resolution as bound states. Thus, they sum all bound state momenta individually in the instant form and obtain the partition function as

$$\mathcal{Z}_{\text{EK}} = \sum_i \sum_{p_i} e^{-\beta\sqrt{M_i^2+p_i^2}} \approx \frac{V}{\pi}\sum_i \int_0^\infty dp_i\, e^{\beta\sqrt{M_i^2+p_i^2}}, \tag{4.95}$$

where the index i lists all Fock states in the chosen K sector, V denotes the spatial volume not to be confused with the light-like compactification length L above. Although all states are discrete at finite K most of the states above the two particle threshold $M \geq 2M_0$ are continuum states, i.e. bound states in relative motion, see Figure 4.1 (A). Taking all these as bound states is conceptually wrong and $\ln \mathcal{Z}_{\text{EK}}$ totally overestimates the correct pressure, even

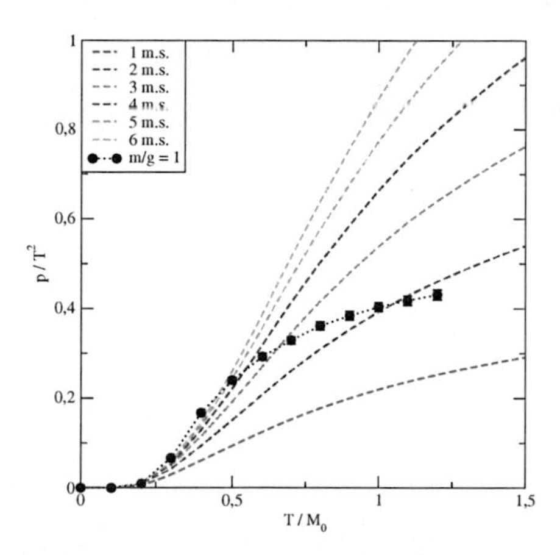

Figure 4.10.: The dimensionless ratio of pressure over temperature squared as a function of temperature. The dashed lines are $p_{\rm mix}/T^2$ as explained in the text for a different number of bound states for the coupling $m/g = 1$ (see legend). The black data points are the same as in Figure 4.8 (A). The dark green curve coincides with the free Bose gas of a particle of mass one.

in the free Fermi or Bose theory. Consequently, the results in [79], like a second order phase transition with critical exponent $\alpha > 0.7$ at about $T/g \approx 1/\sqrt{\pi}$, are not confirmed in our calculation.

Besides the brute-force numerical approach leading to the results above, one may contemplate about more analytical attempts. Since we obtained the spectrum by diagonalizing the LC Hamiltonian one can argue that the eigenstates do not interact and the statistics formulas for the free gas can be used to compute the partition function, the pressure, etc. One way is to imagine the QED_{1+1} gas as a mixture of ideal (Bose) gases of the bound states in the theory. This line of argument leads to the factorization of the total partition function into the partition function of the components, i.e.

$$\mathcal{Z}_{\rm mix} = \prod_{\text{BS } i} \mathcal{Z}_i \tag{4.96}$$

where BS stands for bound states. The total pressure $p_{\rm mix}$ is then the sum of partial pressures. The result for the Schwinger model for $m/g = 1$ is shown in Figure 4.10 and not compatible with 4.8 (A). Clearly the assumption that the continuous spectrum consist simply of bound states in relative motion is misguided and overshoots the full computation.

5. The QCD_{1+1} Equation of state

This chapter gives some perspective to the extension of the results presented in Section 4.4. The idea is to substitute QED_{1+1} by QCD in 1+1 dimensions or, more generally, by fundamental fermions coupled to SU(N) adjoint color gauge fields in 1+1 dimensions. In the light front framework many results for the SU(N) theory can be found in Ref. [31] and the corresponding thesis [101]. The general DLCQ setup applied to QCD_{1+1} is along the algorithm explained in Section 4.2. The gauge field can again be totally eliminated in LC gauge, ignoring the zero mode, by solving the Gauss law for A^-. Thus, one is left with only the fermion field ψ_+ as the independent degrees of freedom. Analogous to the LC Schwinger model Hamiltonian (4.61) the LC QCD Hamiltonian contains terms being combinations of four (anti-)fermion creation and annihilation operators like

$$V_I \sim \frac{g^2}{2\pi} C^{c_1 c_3}_{c_2 c_4} \sum_{n_1,n_2,n_3,n_4} \frac{1}{(n_1+n_3)^2} \delta_{n_1+n_2,n_3+n_4} \, b^\dagger_{n_4,c_4} b^{c_3}_{n_3} d^\dagger_{n_2,c_2} d^{c_1}_{n_1}, \tag{5.1}$$

but includes also SU(N) color factors of the form

$$C^{c_1 c_3}_{c_2 c_4} = [T^a]^{c_1}_{c_4} [T^a]^{c_3}_{c_2} = \frac{1}{2N}\left(N \delta^{c_1}_{c_2} \delta^{c_3}_{c_4} - \delta^{c_3}_{c_2} \delta^{c_1}_{c_4}\right), \tag{5.2}$$

where T^a are the generators of the SU(N) algebra. The SU(N) LF Hamiltonian can be found in Ref. [101]. The computational handling of these theories is more demanding primarily because of the larger state space due to the N fundamental fermions. Compared to the Schwinger model with only electrons and positrons many more partitions of the total momentum are allowed since each part in the partition can be present N times. Although only singlet states need to be considered, because of confinement, the number of states is greater than for the QED_{1+1} interaction. For general SU(N) gauge groups the construction of all color singlet states in terms of the fundamental quarks is a non-trivial problem. The simplest states are single mesons or baryons which can be formed readily out of quarks. All higher representations can be obtained by using the baryons and meson states as building blocks [103]. But this method is redundant in the sense that also linear dependent states are constructed on the way and these should be removed. One suggestion is to orthogonalize the basis states by diagonalizing the Gram matrix, i.e. the matrix obtained by computing the scalar product of all pairs of state vectors. One erases all eigenvectors belonging to a zero eigenvalue since these vectors are actually dependent. For larger harmonic resolutions this procedure will become elaborate because the full spectrum of a large matrix will be necessary. So far results have been reported for rather small resolutions up to $K = 12$ where this computation casts no problems [101].

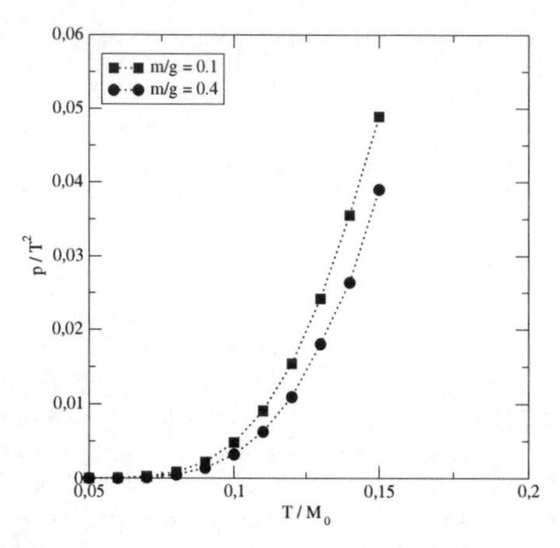

Figure 5.1.: The dimensionless ratio p/T^2 as a function of temperature for QCD_{1+1} at the couplings $m/g = 0.1, 0.4$

Assuming the singlet states have been identified one enjoys another conserved quantum number, namely baryon number B available in these theories. Therefore the Hamiltonian matrix in the free fermion basis is blockdiagonal with respect to B and K. QCD_{1+1} can also be partly 'bosonized' in the strong-coupling limit. Like in the Schwinger model case the fermionic currents are promoted to fundamental fields. Here the Fourier-transformed currents

$$V^a_{k_n} = \frac{1}{2} \int dx^- \; e^{-ik^+_n x^-} j^{+a}(x^-), \tag{5.3}$$

where the color current is $j^{+a} =: \psi^\dagger T^a \psi :$, are used to rewrite the Hamiltonian. The current operators (5.3) satisfy the Kac-Moody algebra. One introduces the momentum-space current V_k belonging to the electro-magnetic current $j^+ =: \psi^\dagger \psi :$ which is sufficient to describe the massless meson states in the strong-coupling limit $m/g = 0$. Besides any massive eigenstate degenerate states can be found by applying V_k. Similar considerations are possible for massless baryonic states [31]. But the full spectrum in the massless limit also contains states which are not generated by the mesonic or baryonic operators.

In view to the thermodynamical application baryon number leads to a baryon chemical potential. Our preliminary result in Fig. 5.1 ignores the baryon chemical potential because we used the $B = 0$ DLCQ matrices up to $K = 12$ generated by Hornbostel/Brodsky[1] [31, 101]. Note the low temperature scale $0 \leq T/M_0 \leq 0.15$ that can be treated within the limited resolution.

[1]private communication with Ken Hornbostel

6. Renormalization group methods

One of the aims of a renormalization program in LC quantization would be to obtain a dynamical mechanism justifying to restrict the particle number in Fock space, i.e. to justify the Tamm-Dancoff approximation for the computation of bound states. This would lead to a rigorous derivation of the phenomenologically successful quark constituent model of hadrons from light cone QCD. Besides the perturbative renormalization group (RG) transformations the standard renormalization scheme proposed for LC quantization is the similarity transformations by Glazek/Wilson [104] or equivalently the flow equations by Wegner [105]. See also [106] for a recent review.

After some general remarks on the renormalization procedure for LC Hamiltonian in the next section we will review some basic features of the similarity transformation approach in the second section. In the third section the density matrix renormalization group (DMRG) is suggested as a non-perturbative numerical implementation of the RG for LC Hamiltonians. The DMRG has its origin in one-dimensional condensed matter systems like the Ising and Heisenberg model and is a generalization of Wilsons numerical renormalization group [107]. A possible algorithm how to apply the DMRG transformation to the massive Schwinger model is outlined.

6.1. Introduction

The key idea of the renormalization group (RG) discovered by Ken Wilson is to absorb high-energy (high-momentum) degrees of freedom into an effective interaction. The transformation is done in such way that the low-energy observables are preserved. Two steps are always included in any implementation of the RG, namely

(i) A blocking transformation to remove high-energy degrees of freedom and lowering the cutoff

(ii) Rescaling of the energy scales and variables by the amount the cutoff was lowered in (i)

The renormalization group is an important concept in high-energy physics (HEP) and in condensed matter physics. In the field of HEP renormalization is usually connected with divergences occurring in perturbative calculations. This would imply that in two-dimensional models no renormalization procedure is needed. Indeed, these models are super-renormalizable and the only divergent diagrams are avoided by normal-ordering. But any regulated quantum field theory is critical (i.e. undergoes a phase transition of second order) once the cutoff is removed. While there is no necessity to apply a renormalization procedure it is certainly desirable to do so. By deriving RG improved Hamiltonians one is able to obtain results closer to the continuum limit at lower values of the cutoff, i.e. with less numerical effort.

This RG idea was adopted to LC quantization [108] and some general remarks are in order. Before one can apply RG transformations to LC Hamiltonians some peculiarities have to be overcome. Local interactions and symmetries are important assumptions for the blocking transformation. On the light cone locality is only present in the transverse direction whereas there are non-local interactions in the longitudinal direction, e.g. the $1/\partial^-$-term is present even in the free LC Hamiltonian. In particular, dimensional counting, valid in the vicinity of a Gaussian fix point, implies different dimensions for longitudinal and transversal momenta. Furthermore a momentum cutoff breaks Lorentz invariance, especially rotational invariance, and often gauge invariance and cluster decomposition is broken by a restriction of particle number. These symmetries/principles ought to be restored in the continuum limit. Counterterms, introduced into the renormalized Hamiltonians by the RG transformation, should remove the symmetry breaking when the renormalization procedure is completed. Generally, the number and structure of the counterterms are restricted by symmetry considerations and it is not surprising that many operators are allowed in the LC Hamiltonian since continuous symmetries are broken by the LC setup. It is useful to separate longitudinal and transverse renormalization group transformations because one observes that light cone Hamiltonians are invariant under scaling in longitudinal direction because of the boost-invariance of the longitudinal momentum fractions $x = k^+/P^+$. This has been appreciated in the simple scaling law in L of the Schwinger model light cone Hamiltonian in Section 4.2. Under transversal RG transformation (relevant, marginal, and irrelevant) counterterms are generated depending on entire functions of longitudinal momentum. On the other hand rescaling of the longitudinal momentum introduces counterterms depending on functions of the transverse momenta. The immediate question arises whether these operators can be ultimately fixed by a finite number of phenomenological inputs, like physical masses or couplings. Since infinitely many conditions must be fulfilled it seems that one looses renormalizability for theories where it was established in the instant form. This problem was solved for simple theories by invoking a concept called coupling constant coherence. Hereby not all couplings are independent functions, but rather a to be defined finite subset of the couplings determine the remaining ones.

6.2. Similarity transformation renormalization group

Let us assume that the physical system under consideration is given by some Hamiltonian matrix. The Hamiltonian is divided into $H = H_0 + H_I$ where H_0 is a well-controlled operator that one is able to diagonalize. We take the eigenvectors of H_0 as the basis of the full Hamiltonian and sort the states by increasing free energy. The idea is to apply continuous unitary transformations to eliminate off-diagonal matrix elements. The off-diagonal elements are representing transitions between states which have large difference in energy. Therefore the flow-parameter Λ is introduced and the transformed Hamiltonian is

$$H(\Lambda') = U^\dagger(\Lambda, \Lambda') H(\Lambda) U(\Lambda, \Lambda'), \tag{6.1}$$

where $H(\Lambda)$ denotes the Hamiltonian at the cutoff scale Λ. For $\Lambda \to 0$ the Hamiltonian $H(\Lambda)$ converges to the diagonal form. The $U(\Lambda)$ are called similarity renormalization group (SRG) transformations. Reducing (6.1) to the infinitesimal transformations lead to the flow equations

$$\frac{d}{d\Lambda} H(\Lambda) = [\eta(\Lambda), H(\Lambda)], \tag{6.2}$$

with the anti-hermitian generator

$$\eta(\Lambda) = \frac{dU(\Lambda)}{\Lambda} U^\dagger(\Lambda) = -\eta^\dagger(\Lambda). \tag{6.3}$$

There is the freedom to chose the generator aiming at a certain property the transformed Hamiltonian. One common choice is the generator

$$\eta(\Lambda) = [H_d(\Lambda), H(\Lambda)], \tag{6.4}$$

where H_d is the diagonal part of H. Another option is to try to bring the Hamiltonian in block-diagonal form instead of full diagonalization. Thereby particle number changing terms in the Hamiltonian are eliminated by the unitary transformations [106]. The formal solution of (6.2) reads

$$H(\Lambda) = U^\dagger(\Lambda) H U(\Lambda) \quad \text{with} \quad U(\Lambda) = T_\Lambda \, e^{\int_0^\Lambda d\lambda\, \eta(\lambda)}, \tag{6.5}$$

where the ordering symbol T_Λ with respect to Λ is introduced because the generators in general do not commute for different Λ.

From the DLCQ point of view the SRG does not try to integrate out longitudinal high-momentum modes or to address the limit $K \to \infty$. It rather concerns the energy differences between states and restricts the bandwidth of the Hamiltonian matrix. However, the SRG suggests a way to find counterterms which are thought to reduce also the dependence of low-energy observables on the momentum cutoff K [104]. It also solves convergence problems if perturbation theory in the interaction Hamiltonian H_I is used to compute low-energy observables.

Many applications of the SRG to light front field theory have been given. A few examples are mentioned here. In [109] the matrix elements of a renormalized Hamiltonian for the asymptotic free scalar ϕ^3 theory in six dimensions were computed in perturbation theory up the third order in the (scale dependent) coupling. Heavy quarkonia states were addressed in [110] and an effective Hamiltonian is obtained from QCD to second order in the coupling with a bandwidth representing the hadronic scale. The effective Hamiltonian is divided into a valence sector, including a logarithmic confinement and Coulomb potential treated non-perturbatively (i.e. by diagonalization). The remaining interaction is subject to perturbation theory. Low-lying masses of glueballs were computed using the SRG for the pure-glue QCD Hamiltonian in [111]. Besides the perturbative expansion of the renormalized interaction to second order a Tamm-Dancoff cut to two-particle states was enforced. Still the obtained glueball spectrum is in fairly good agreement with the results of other methods (see Table I in [111]). Numerically asymptotic freedom of the pure gauge theory was confirmed but the couplings was found to vanish faster as the standard result $\alpha_s(\Lambda) \sim 1/\ln\Lambda$ as $\Lambda \to \infty$. This effect may be due to the Tamm-Dancoff cut. Finally, the formalism of flow equations was utilized in [112] to calculate the positronium spectrum. To this end the flow equation, again expanded to second order, was solved and the effective Hamiltonian was truncated to the two-particle sector.

A common theme of all publications is the space of renormalized Hamiltonians being fixed by perturbative expansions from the very beginning. Thus, the generator $\eta(\Lambda)$ is constructed to a given order in the renormalized canonical coupling. This should suffice for a survey on the SRG and we focus on yet another implementation of the renormalization group.

6.3. Density matrix renormalization group and DLCQ

General setup

The density matrix renormalization group (DMRG) [113, 114] is regarded as the most successful non-perturbative real-space realization of the renormalization group idea. In a transparent way DMRG builds on Kadanoffs blocking transformation, thinning out of irrelevant degrees of freedom, and deriving effective interactions. For local interactions, like the one-dimensional Heisenberg model for spin chains, numerical exact results with a relative accuracy of 10^{-9} for the ground state energy can be reached with modest computational effort. The application of DMRG to LC Hamiltonians was suggested in [115] after showing the efficiency in a non-perturbative quantum mechanical toy model usually considered in the similarity renormalization scheme by Glazek/Wilson. The original DMRG method is formulated in the language of spin chains. For the introduction we will use the somewhat unfamiliar spin chain terminology, later the contact to DLCQ will be established.

The DMRG algorithm tries to approximate the ground state wavefunction (or few low lying target states) by only a small number of states compared to the whole state space at a certain chain length. To this end the chain is divided into a system and an environment part and the union of both is called the superblock. The ground state of the superblock has the expansion

$$|\Psi\rangle = \sum_{m^S,\sigma^S} \sum_{m^E,\sigma^E} \psi_{m^S\sigma^S m^E\sigma^E} |m^S\sigma^S\rangle |m^E\sigma^E\rangle, \tag{6.6}$$

where m^S (m^E) are the system (environment) degrees of freedom of the old block and σ^S (σ^E) the degrees of freedom of a exclusive lattice site. One derives the density matrices of the system and the environment blocks by tracing out the opposite block, i.e.

$$\rho_S = \mathrm{Tr}_E\,\rho, \qquad \rho_E = \mathrm{Tr}_S\,\rho, \tag{6.7}$$

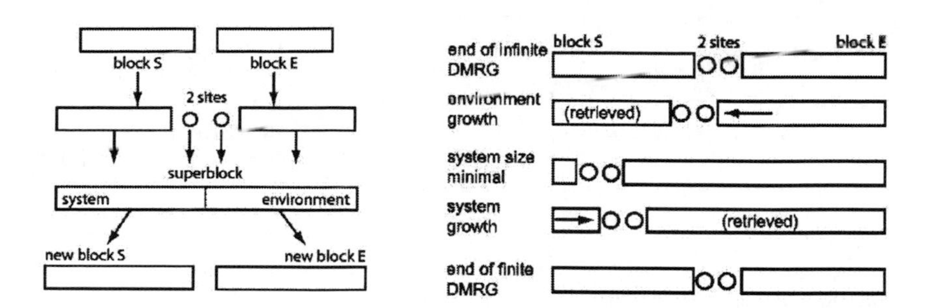

Figure 6.1.: Illustration of the DMRG algorithms. (Left panel) The infinite size DMRG algorithm. The system (S) and the environment (E) block grow by a lattice size each. (Right panel) In the finite system DMRG the system block grows while the environment block shrinks and vice versa. Both pictures taken from [116].

with the density matrix of the pure state $\rho = |\Psi\rangle\langle\Psi|$. The reduced density matrices are then diagonalized by solving

$$\rho_{S/E}\,|\omega_\alpha^{S/E}\rangle = \omega_\alpha^{S/E}|\omega_\alpha^{S/E}\rangle, \tag{6.8}$$

and the eigenstates of large statistical weight $\omega_\alpha^{S/E}$ are stored. This way the most probable states of the system or environment block contributing to the ground state of the superblock are kept. Furthermore the subblocks are not in a pure state but rather described by density matrices and the entanglement of the system with the statistical bath is accounted for. The number N of retained states is a variational parameter and the value can be chosen on computational demands. Although the DMRG aims to compute the ground state it can be viewed as a renormalization group transformation to obtain effective Hamiltonians matrices which have the same ground state. Further characterizations are necessary to integrate DMRG into the standard notions of RG like marginal and (ir-)relevant operators, beta functions for the couplings etc.

The density matrices are normalized to one but the truncation introduces a deviation, namely the truncation error

$$\varepsilon_N = 1 - \sum_{i=1}^{N} \omega_i^S. \tag{6.9}$$

For gapped systems the truncation error is expected to decrease exponentially with increasing number of retained states.

As a next step one enlarges the spin blocks and monitors how the retained states of the blocks change under the running of the DMRG. There exist two versions of the algorithm: the infinite system and the finite system approach.

First we review the infinite size version. Here the total length of the chain is not fixed and each block is enlarged by one lattice node. The setup is shown in Figure 6.1 left panel.

After the truncation of the system and environment blocks, a new superblock is formed by the retained states and the two added nodes. Usually the infinite algorithm is implemented as a warm-up step to created the initial system and environment blocks for the different parts of the chain. Further accuracy is gained by running the finite size algorithm. Here the size of the chain is fixed, commonly at the maximal length at the end of the infinite algorithm. The system block is now growing in expense of the environmental block and both are updated in every step. When one block reaches the maximum size the growing direction is altered, see Figure 6.1 right panel. A DMRG sweep is finished when the initial block configuration is reached again.

Different suggestions, where the two lattice sites are best added, are found in the literature depending on the boundary conditions [116]. More variations of DMRG exists, say, with only one site insertion [117] which is in the simplest implementation less accurate than the standard version. An active area of research is the extension of DMRG to higher dimensional systems. A simple idea is to single out one direction for the renormalization process under the assumption that the other directions are not or weakly interacting. The observation that DMRG is a variational approach on the polynomially sized subspace of matrix product states, see the Ref. in [116], leads to an abundance of DMRG-like algorithms using a different sets of states, e.g. [118]. As a first attempt we focus on the standard algorithm in one-dimensional momentum space.

The usefulness of the DMRG in the context of quantum field theoretical systems may be questioned because of the infinite degrees of freedom at each lattice site and the in general non-local interactions. However, some successful instant form applications are given in [80, 119]. In the first Ref. [80] a DMRG implementation for the lattice Hamiltonian (4.75) was given and the accuracy of the estimates of the lowest mass eigenstate were increased by an order of magnitude. In the latter the scalar ϕ^4 theory is treated, thereby estimating the critical coupling of spontaneous parity breaking very accurately. Carrying the DMRG over to DLCQ Hamiltonian needs a formulation in momentum space. This has been done in [120, 121] for condensed matter systems trying to utilize the total momentum as an additional conserved quantum number. One immediate drawback in the momentum space DMRG is that only the finite size algorithm is applicable. Instead of the infinite size algorithm the authors of [120] suggested to use the conventional numerical renormalization group [107], i.e. keeping only the lowest energy states in the expansion of the ground state, for the warm-up phase. Furthermore the transformation to momentum space leads to non-local interactions, even if the interaction was local in real-space, which are harder to deal with for the DMRG .

Application to DLCQ

Now we will outline the application of momentum space DMRG to the DLCQ Hamiltonian of the Schwinger model. The key idea is to look at the single-particle fermion momentum mode as sites of the 'chain' in LC momentum space. The possible states at an arbitrary node of the chain are the following four: $\{|0\rangle, b_n^\dagger|0\rangle, d_n^\dagger|0\rangle, d_n^\dagger b_n^\dagger|0\rangle\}$. The conventional DLCQ does not fix the number of momentum modes but the total momentum K. It is not clear how to implement a growing step $K \rightarrow K+1$ in the standard DLCQ and how to use the informations about the 'smaller' system since the partitions of K and $K+1$ are not obviously related. The picture of the momentum chain makes the growing step easy and in every step the state space is enlarged by a factor of four where one part simply consists of all former configurations. One faces the fact that several K sectors have to be accounted for simultaneously but at least the LC Hamiltonian is block-diagonal in K. Another challenge to be resolved is the separation of the chain into the system and environment blocks. For local interactions, connecting neighboring lattice points, a left-right separation, as shown in Figure 6.1, is efficient because the blocks only interact at their boundaries. This is not the case in momentum space and algorithms have been suggested [116] how to choose the blocks based on entanglement entropy measures. Discussing these would go beyond the scope of this thesis. For simplicity we will consider in the following a left-right separation of the system and environment block. As in the standard DLCQ one has to determine all possible partitions and thus Fock vectors for the chain. The allowed resolutions span from the lower end of the chain $K_{\text{min}} = k_l$ to $K_{\text{max}} = \frac{1}{2}\left(k_h(k_h+1) - k_l(k_l-1)\right)$, where $k_h(k_l)$ is the largest (smallest) one-particle momentum. The Fock vectors can be computed during the growing step but in doing so one has to store also states with $Q \neq 0$ since these can contribute to physical states (i.e. $Q|phys\rangle = 0$) in the later process. In the next step the initial blocks are computed. For small systems the corresponding Hamiltonian is exactly diagonalized, but with increasing chain length the high LC energy states are neglected. Nevertheless, few states in every (K, Q) sector should be saved to ensure all (K, Q) are explored later. The (partial) basis transformation is also stored. When a superblock is formed during the finite system DMRG algorithm reordering of the Hamiltonian may be necessary to restore the block-diagonal structure in K. By reordering the clear separation of system and environment states gets lost and one should take care to identify which states to trace out in (6.7).

It will be very interesting if the DMRG is able to deliver consistent results for the long-ranged (in real-space) Coulomb potential in the LC Schwinger model. The naive exact diagonalization approach to the LC chiral Schwinger model already leads to accurate results. We expect DMRG to be able to handle resolutions not reached in any DLCQ calculation before and to gain orders of magnitude more accurate mass estimates than given in Table 4.1.

When the LC wave function of the ground state is determined one can immediately compute observables $\mathcal{O}$ by $\langle\Psi|\mathcal{O}|\Psi\rangle$. But this may not be the optimal choice for some observables. In some cases it is more effective to apply the DMRG algorithm directly to the observable under consideration. One of the possible scenarios is the usage of DMRG for quantum systems at finite temperature. Here one does not retain states according to the Hamiltonian but instead according to the transfer matrix

$$T = e^{-\varepsilon H}. \tag{6.10}$$

After its main object of interest this method is called (with some notational abuse) the transfer matrix renormalization group (TMRG) [122]. The crucial step for this method is the effective determination the transfer matrix operator T. In case of the local spin chain Hamiltonians containing one-site interactions only, it is possible to approximate partition function by local transfer matrices by staggering;

$$\mathcal{Z} = \lim_{M\to\infty} \mathrm{Tr}\left[e^{-\frac{\beta}{M}H_o}e^{-\frac{\beta}{M}H_e}\right]^M = \lim_{M\to\infty} \mathrm{Tr}\, T_M^{N/2}, \tag{6.11}$$

where H_o, H_e are the local Hamiltonians coming from odd or even lattice sites, M and N denote the Trotter number and the chain length correspondingly. The local transfer matrices T_M can be expressed as simple matrix products of local spin states. In the thermodynamical limit $N \to \infty$ only the largest eigenvalue λ of T_M contributes. The TMRG is set up to approximate λ for increasing Trotter number M and left and right eigenvector of the (generally asymmetric) matrix T_M. For small values of M accurate thermodynamical results are obtained in the high-temperature regime.

There are obstacles in utilizing this algorithm for the DLCQ Hamiltonian of the massive Schwinger model. Most importantly the decomposition (6.11) is not possible because of the non-local coupling between lattice sites in momentum space. Opposed to the spin chain models in real-space the LC Hamiltonians are not simple sums of local Hamiltonians. Therefore finding a decomposition in analogy into 'odd' and 'even' lattice sites such that the transfer matrices decouple is a hard (if not impossible) task.

In conclusion, we have described a possible way to use the DMRG in a field theory regularized by DLCQ. Since the Hamiltonians of spin chain models in momentum space are relatively similar to the DLCQ Hamiltonian of one-dimensional field theories one is able to translate the DMRG algorithm. But the state space has to be constructed in a different fashion compared to the regular DLCQ approach. Also the cutoff is not the total momentum P^+ (as in DLCQ) but the largest possible one-particle momentum, corresponding to the length of the spin chain. How the system and environment blocks should be separated is open for optimization and in the described algorithm a simple left-right separation was chosen. Sorting the contributing states at each chain length by their total $+$ momentum restores the block-diagonal structure of the LC Hamiltonian. In the massive Schwinger model the low-lying bound states can be accurately expanded in free few-particle states which is only a polynomial increasing set of states. Therefore we expected the DMRG algorithm, that also cuts the Fock space to a relevant polynomial set, to perform very well. Numerical results will be reported in subsequent publications.

7. Conclusions and Perspectives

In this thesis the thermodynamical properties of LF quantized QED_{1+1} and QCD_{1+1} have been presented. The light front gauge theories have been regularized by DLCQ. The formulation of thermal LC field theory is subtle because temperature is usually identified with the compactification length of a time-like direction. In LC quantization the time evolution parameter is light-like and a naive generalization runs into problems. Applying the general LF formalism one is able to derive a proper statistical operator. We have used this operator to compute the pressure, energy and entropy density in the thermodynamical limit. Taking the thermodynamical limit is crucial to obtain unambiguous results not depending on the artificial light-like box. Unfortunately, going to large box sizes requires to work with high harmonic resolutions or, technically stated, with large matrices representing the LC Hamiltonian.

The program code for DLCQ QED_{1+1} has been rewritten to handle harmonic resolutions up to $K = 54$. For comparison, older programs operated at $K = 20$ [79]. The large resolution allowed us to compute thermodynamical observables in the temperature interval $0 \leq T/M_0 \leq 1.5$, where M_0 is the ground state mass. These results are state-of-the-art for the massive Schwinger model as no comparable results from other non-perturbative techniques, like finite temperature lattice gauge theory, are available. By altering the coupling in the massive Schwinger model we investigated the transition from the free fermion gas to a non-interacting Bose gas. Former investigations [79] found a second order phase transition but these observations have not been confirmed. The numerical methods have been tested for the ideal fermionic gas in the grand-canonical ensemble.

For QCD_{1+1} we have used the DLCQ matrices of Hornbostel/Brodsky with resolutions $K \leq 12$. Due to the small resolutions we were limited to the temperature range $T/M_0 \leq 0.15$. Nevertheless, we like to stress that these are the first results of the LC QCD_{1+1} pressure.

The improved harmonic resolution was also used to enhance the accuracy of typical observables, like bound state masses and structure functions, at zero temperature. For certain couplings two orders of magnitude more accurate bound state masses have been obtained. The results are highly competitive to the ones given by lattice Hamiltonian computations. Further accuracy could be gained by applying renormalization group algorithms, like the density matrix renormalization group (DMRG), to the massive Schwinger model. We have described the translation of the DMRG from spin chain models to DLCQ Hamiltonians in detail, pointing out the technical challenges. The topological properties of QED_{1+1} have been investigated in the LF quantized version. Thereby, we have explained the realization of the θ-vacua as coherent states of the gauge field zero mode and fermionic excitations.

There are many possible avenues to go beyond the presented work, a list is given below.

We begin the discussion with the most obvious suggestions. It is clear that invoking more computational power, e.g. running the programs on super-computers instead of work stations, would lead to an increase in harmonic resolution. A higher resolution results of course in more accurate mass eigenvalues and wavefunctions. In the thermodynamical application the temperature range could be enlarged and the errors could be reduced. Especially the QCD_{1+1} code from [31] could use some renewal. A next step would be to compute the DLCQ approximation of the spectral function evaluating (3.47), thereby determining the thermal mass and widths of the bound states. For the thermal mass of the Schwinger boson a perturbative prediction exists [123].

The critical properties of the massive Schwinger model at $\theta = \pi$ can be accessed by an implementation of the background electric field term in the LC Hamiltonian. Following [124] the additional term reads

$$P_\theta^- = -\frac{g^2\theta}{2\pi}\int dx^- j^+(x^-)x^- . \tag{7.1}$$

Such a background term would allow an independent check of the critical coupling estimate (4.8) by LC methods. Note that the θ-term is a complex contribution to the Schwinger model Hamiltonian (4.62) that breaks momentum conservation. In principle, the vacuum periodicity should manifest itself, e.g., in a theta dependent Schwinger boson mass.

On the formal side, the theoretical development of thermo field dynamics, a Hamiltonian-based formulation of thermal field theories [69], in the (general) light front frame is still missing. Because of the intrinsic Hamiltonian nature of the LC quantization it appears, from a theoretical point of view, natural to seek the generalization to finite temperature in this framework. Again the question of the simplicity of the LC vacuum and zero modes have to be addressed since the thermal ground state within thermo field dynamics is given by a Bogoliubov transformation of the zero temperature vacuum.

Yet another direction is the task to extend the results of this work to four dimensions. As pointed out in Section 2.3 two possibilities have been developed for four-dimensional gauge theories: the transverse lattice and the basis function approach. For a recent suggestion to utilize the harmonic oscillator wave functions for the transverse degrees of freedom, see [125]. Since both approaches derive a LC Hamiltonian matrix one is,in principle, able to apply the methods outlined in this work to compute the partition function. However, there is also an alternative approach, the gauge-invariant Schrödinger representation introduced in [126]. We suggest to use the so-called KKN variables instead of gauge-links in the transverse directions and hence achieve a different transverse lattice gauge theory.

The renormalization procedure for LC Hamiltonians is spoiled by the explicit breaking of Lorentz-symmetry and the corresponding unrestrainedness of counterterms. The concept of coupling constant coherence was suggested to cure this problem in renormalizable theories. A possibly different angle to attack this problem is to utilize the de Donder-Weyl formalism of covariant Hamiltonian dynamics, see e.g. [127]. One introduces canonical momenta not only for the time derivative of fields but also for the spatial ones and thus treats time and space on an equal footing. This leads by Legendre transformation to a Lorentz-invariant Hamiltonian. The usefulness of this idea in singular LC systems has to be shown in future work.

A. Notation

Light Front Coordinates

- light cone coordinates in Lepage-Brodsky (LB) notation

$$\begin{aligned} x^+ &= x^0 + x^3, \\ x^- &= x^0 - x^3, \\ x_\perp^i &= x^i, \qquad i = 1, 2. \end{aligned} \tag{A.1}$$

- light cone metric in LB

$$\eta_{\mu\nu} = \begin{pmatrix} 0 & 1/2 & 0 & 0 \\ 1/2 & 0 & 0 & 0 \\ 0 & 0 & -1 & 0 \\ 0 & 0 & 0 & -1 \end{pmatrix}, \qquad \eta^{\mu\nu} = \begin{pmatrix} 0 & 2 & 0 & 0 \\ 2 & 0 & 0 & 0 \\ 0 & 0 & -1 & 0 \\ 0 & 0 & 0 & -1 \end{pmatrix}. \tag{A.2}$$

- Minkowski product

$$a \cdot b = 2a_+b_- + 2a_-b_+ - a_ib_i = \frac{1}{2}a^+b^- + \frac{1}{2}a^-b^+ - a^ib^i \tag{A.3}$$

- partial derivatives

$$\begin{aligned} \partial^- &= \frac{\partial}{\partial x_-} = 2\frac{\partial}{\partial x^+} = 2\partial_+ \\ \partial^+ &= \frac{\partial}{\partial x_+} = 2\frac{\partial}{\partial x^-} = 2\partial_- \end{aligned} \tag{A.4}$$

$$A\overleftrightarrow{\partial_\pm}B = A(\partial_\pm B) - (\partial_\pm A)B \tag{A.5}$$

Oblique Light Front Coordinates

- space-time

$$\begin{aligned} x^{\bar{0}} &= x^0 + x^3, \\ x^{\bar{i}} &= x^i, \quad i = 1, 2, 3 \end{aligned} \tag{A.6}$$

- momentum space

$$
\begin{aligned}
p_{\bar{0}} &= p_0 = p^0, \\
p_{\bar{3}} &= -p_0 + p_3 = -p_- = -\frac{1}{2}p^+, \\
p_{\bar{i}} &= p_i, \quad i = 1,2
\end{aligned}
\tag{A.7}
$$

Representation of the Gamma Matrices

- explicit representation of Dirac matrices

$$
\gamma^0 = \begin{pmatrix} 0 & \mathbf{1} \\ \mathbf{1} & 0 \end{pmatrix}, \qquad \gamma^3 = \begin{pmatrix} 0 & -\mathbf{1} \\ \mathbf{1} & 0 \end{pmatrix}, \qquad \gamma^j = \begin{pmatrix} -i\sigma^j & 0 \\ 0 & i\sigma^j \end{pmatrix}, \tag{A.8}
$$

the elements are 2×2 matrices and $\sigma^j,\ j = 1,2$ are the Pauli spin matrices

- explicit representation of the projection operators

$$
\Lambda^+ = \frac{1}{4}\gamma^-\gamma^+ = \begin{pmatrix} \mathbf{1} & 0 \\ 0 & 0 \end{pmatrix}, \qquad \Lambda^- = \frac{1}{4}\gamma^+\gamma^- = \begin{pmatrix} 0 & 0 \\ 0 & \mathbf{1} \end{pmatrix} \tag{A.9}
$$

Poincaré Generators

- energy-momentum tensor

$$
T^{\mu\nu} = \frac{\partial \mathcal{L}}{\partial\left(\partial_\mu \phi_r\right)} \partial^\nu \phi_r - g^{\mu\nu}\mathcal{L} \tag{A.10}
$$

- angular momentum-boost tensor

$$
M^{\mu\nu} = \frac{1}{2}\int dx^- d^2x^\perp \left\{ x^\mu T^{+\nu} - x^\nu T^{+\mu} + \frac{\partial \mathcal{L}}{\partial\left(\partial_+ \phi_r\right)} \Sigma_{rs}^{\mu\nu} \phi_s(x) \right\}, \tag{A.11}
$$

where the second term is related to the spin of the fields $\phi_r(x)$

$$
\Sigma_{rs}^{\mu\nu} = \begin{cases} 0, & \text{scalar} \\ \frac{1}{4}\left[\gamma^\mu, \gamma^\nu\right]_{rs}, & \text{fermion} \\ \eta_r^\mu \eta_s^\nu - \eta_s^\mu \eta_r^\nu, & \text{vector} \end{cases} \tag{A.12}
$$

- momenta

$$
P^\mu = \frac{1}{2}\int dx^- d^2x^\perp T^{+\mu} \tag{A.13}
$$

B. Special functions

During the derivation of the LC Schwinger model Hamiltonian several special functions have been used following Ref. [88]. The Green functions of the constraints (4.25), (4.26) have to be carefully defined on the finite interval $\left[-\frac{L}{2}, \frac{L}{2}\right]$. Two points are important: first, in case one deals with equations only containing normal mode quantities the zero mode of the delta distribution should be subtracted. Second, the Green function should obey the same boundary conditions as the functions in the original equation. Note that the $n = 0$ contribution is excluded from the sums below.

- discrete delta distribution

$$\delta_L(x) = \frac{1}{L} + \frac{1}{L} \sum_{n=-\infty}^{\infty} e^{i\frac{2\pi}{L}\pi x} = \frac{1}{L} + \mathcal{D}_L(x) \tag{B.1}$$

- discrete delta distribution without zero mode

$$\mathcal{D}_L(x) = \delta_L(x) - \frac{1}{L} \tag{B.2}$$

- discrete sign distribution - anti-periodic

$$\varepsilon_L(x) = \frac{x}{L} + \frac{1}{2i\pi} \sum_{n=-\infty}^{\infty} \frac{1}{n} e^{i\frac{2\pi}{L}\pi x} + c \tag{B.3}$$

- discrete sign distribution without zero mode - periodic

$$\mathcal{E}_L(x) = \varepsilon_L(x) - \frac{x}{L} + c \tag{B.4}$$

- discrete modulus function

$$\xi_L(x) = \frac{x^2}{2L} - \frac{L}{4\pi^2} \sum_{n=-\infty}^{\infty} \frac{1}{n^2} e^{i\frac{2\pi}{L}\pi x} + cx + d \tag{B.5}$$

- discrete modules function without zero mode

$$\Xi_L(x) = \xi_L(x) - \frac{x^2}{2L} - \frac{L}{6} + cx = \frac{1}{2}|x| - \frac{x^2}{2L} - \frac{L}{6} + cx \tag{B.6}$$

The generalized functions $\varepsilon_L(x)$ and $\xi_L(x)$ are obtained by integration of (B.1). The integration constant c can be set to zero. In Eq. (B.6) the constant d was chosen such that $\int \Xi_L(x) = 0$ holds.

C. Interaction vertices of the massive Schwinger model

For the perturbative expansion the diagrammatic representation of the LC Hamiltonian is very useful. Although in this work no perturbative computation were carried out we feel that the diagrams highlight the physical meaning of each term occurring in the interaction part of the LC Hamiltonian (4.62). The diagrams below are of course *no* Feynman graphs met in covariant perturbation theory nor diagrams from (old-fashioned) Hamiltonian i.e. time-ordered perturbation theory. The wavy line stands for the instantaneous photon propagator and the solid lines represent the particles or antiparticles according to the direction of the arrows. By definition the particle direction is from left to right.

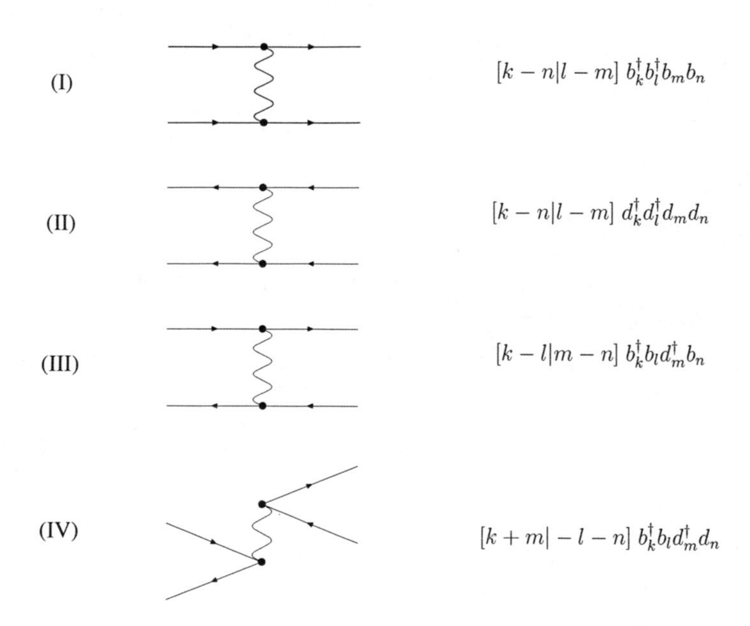

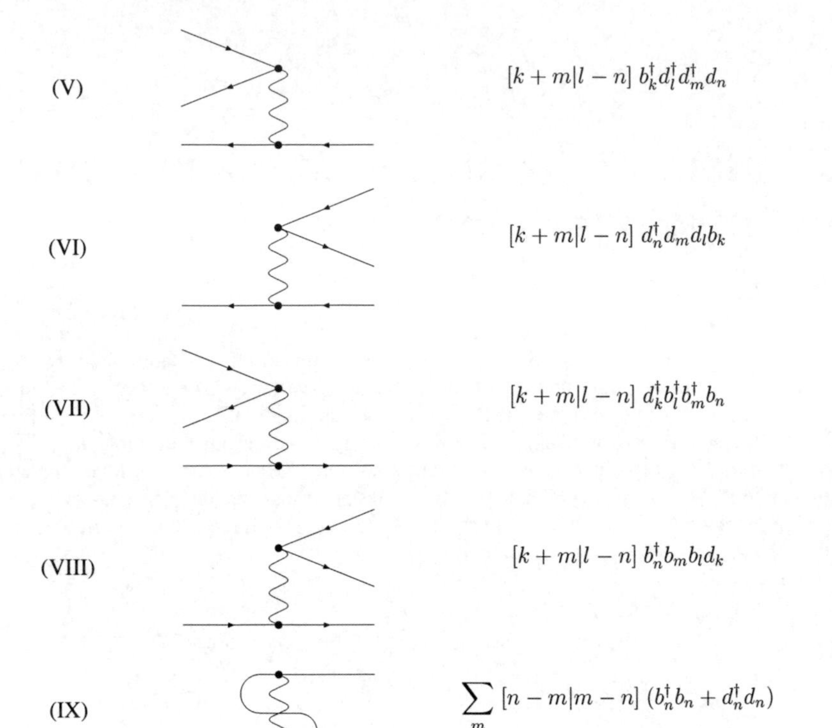

(X)

$$\sum_m [n+m|-n-m]\ (b_n^\dagger b_n + d_n^\dagger d_n)$$

D. Technical details

We like to give some details on how the presented DLCQ computation was practically implemented. Writing a DLCQ program from scratch one first of all has to construct the Fock state space, secondly, one computes all operators one is interested in as matrices on different Fock space sectors. If the mass spectrum of the theory (or some part) is a quantity of interest the diagonalization of matrices will be necessary. Finally since our goal here is to compute thermodynamical properties of QED_{1+1} we calculate the trace of matrix exponentials.

D.1. Integer partitions and Fock space

Former DLCQ computations, e.g. [101, 30, 128, 32], have used a rather small harmonic resolution K or explicit bounds on the number of parts within each partition because of the limited computer power. In that case it is clear how to find all possible partitions by direct search. If all integer partitions should be found and the resolution is large a direct search is very inefficient. We have developed a recursive algorithm that will be outlined in the following.

Given an integer $K \in \mathbb{N}$ we want to construct all partitions with each part being of N kinds in general. This set of partitions would correspond to a state space for N kinds of fermions and its cardinality will be denoted by $p_N(K)$. Because of the Pauli principle no $N+1$ fermions can occupy the same one-particle state. Formally in the limit $N \to \infty$ one obtains the prescription for the boson state space. The key idea is to consider strictly ordered partitions, that means the parts are increasing from left to right, and construct these sorted by the number of parts. Practically, we first construct all partitions with each part being repeated maximally N times. Afterwards the degeneracy of each part into N kinds will be looked at. We call partitions consisting of i parts i-partitions.

The algorithm starts trivially with 1-partitions which is only K itself. Then continues with sorted 2-partitions which range for $(1, K-1)$ to $(K/2, K/2)$ or $([K/2], [K/2]+1)$ depending if K is even or odd. Note that $(K/2, K/2)$ is not allowed if $N = 1$. Here $[K/2]$ stands for the largest integer less than $K/2$. Progressing further to 3-partitions we insert for the last part of each 2-partition all possible ordered 2-partition of that particular number. These partitions are stored on hard disk if one has already computed the partitions for smaller K before. The new 3-partitions have to fulfill two requirements: first they have to be compatible with the Pauli principle above and second they have to be ordered. From this point the algorithm proceeds to 4-partitions and so on until all ordered partitions are compiled. The longest and therefore last partition has the structure

$$(\underbrace{1,\ldots,1}_{N\text{ times}},\underbrace{2,\ldots,2}_{N\text{ times}},\ldots). \tag{D.1}$$

As an example we consider $K = 10$ and $N = 2$:

$$\begin{aligned}
\text{1-partitions} &: \quad (10) \\
\text{2-partitions} &: \quad (1,9),(2,8),(3,7),(4,6),(5,5) \\
\text{3-partitions} &: \quad (1,1,8),(1,2,7),(1,3,6),(1,4,5) \\
&\quad\;\; (2,2,6),(2,3,5),(2,4,4) \\
&\quad\;\; (3,3,4) \\
\text{4-partitions} &: \quad (1,1,2,6),(1,1,3,5),(1,1,4,4) \\
&\quad\;\; (1,2,2,5),(1,2,3,4),(2,2,3,3) \\
\text{5-partitions} &: \quad (1,1,2,2,4),(1,1,2,3,3)
\end{aligned}$$

Fock states are derived from the partitions by observing that one partition can be mapped to multiple Fock state since one can assign the one-particle momentum to each particle kind. For the generic case of N kinds of particles this leads to the following degeneracy

$$d = \prod_i \binom{N}{n_i}, \tag{D.2}$$

where $0 \leq n_i \leq N$ counts how often k_i occurs in $\sum k_i = K$.

The sequence $p_N(K)$ is given by the expansion coefficients of

$$f(x) = \prod_{m=1}^{\infty} (1 + x^m)^N \tag{D.3}$$

into a power function in x. Furthermore $p_N(K)$ can be obtained by Euler transformation (in K) of the period 2 sequence

$$[N, 0, N, 0, \ldots]. \tag{D.4}$$

In Figure D.1 are various sequences $p_N(K)$ for small N shown. Since the ordinate in D.1 has a logarithmic scale one concludes that $p_N(K)$ is exponentially growing in K. This is not unexpected because the Hardy-Ramanujan asymptotic expression for partitions of an integer n without any restrictions (in physics language for boson states) reads

$$p(n) \sim \frac{\exp\left(\pi\sqrt{2n/3}\right)}{4\sqrt{3}n} \text{ as } n \to \infty. \tag{D.5}$$

The exact formula for the number of partitions of an integer n, the so-called Rademacher formula, is known but a rather complicated expression. It can be found in textbooks on partitions like [129].

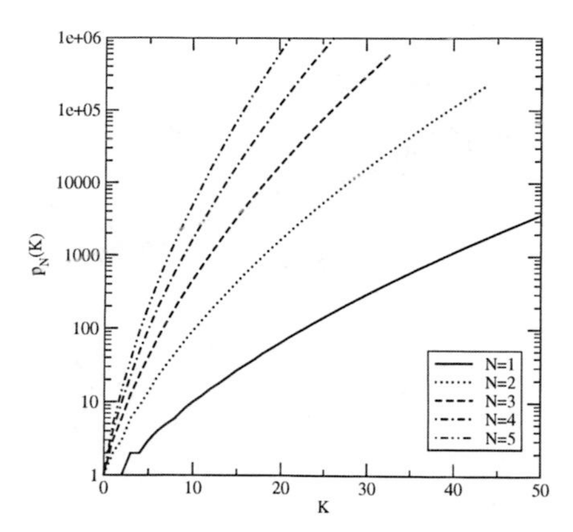

Figure D.1.: The number of partitions with maximally N times the same number as a function of the harmonic resolution. Note the logarithmic scale of the y-axis.

D.2. Matrix representation and diagonalization of the LC Hamiltonian (4.61)

The problem of dealing with sparse matrices of large dimensionality is common in science and powerful methods have been developed. Considering the LC Hamiltonian (4.61) with the interactions (4.62) many matrix elements vanish because of the momentum conservation. Furthermore (4.62) only contains terms which conserve particle number or change it by two. Remember that the number of particles is always even because of confinement in the Schwinger model.

In practice we sorted the partitions/Fock states for each resolution K by increasing particle number and ensured the vanishing of the matrix elements far away from the diagonal. Figure D.2 shows the matrix structure for the harmonic resolution $K = 40$. An intriguing repetitive pattern is visible, but this argument could not be made strict. From a numerical view point matrices with small bandwidth are better conditioned. Observing the large void areas in Fig. D.2 one may wonder whether any ordering of lines/rows exists for which these spaces are filled. Having a small upper bound on the matrix bandwidth would offer a major computational advantage. Generic algorithms like the Cuthill-McKee algorithm are suitable but we found that the results are rather discouraging for the LC Hamiltonian matrices without being able to state the exact reason for the failure.

Nevertheless, the matrix density, the ratio of non-vanishing matrix elements to the size of the matrix, is of the order of 10^{-3}. The matrix density, plotted in Figure D.3 for various resolutions, decreases for large K. This effect is surely desirable since the matrices are stored in sparse format on hard disk. We do not discuss the utilized sparse routines in detail here, but

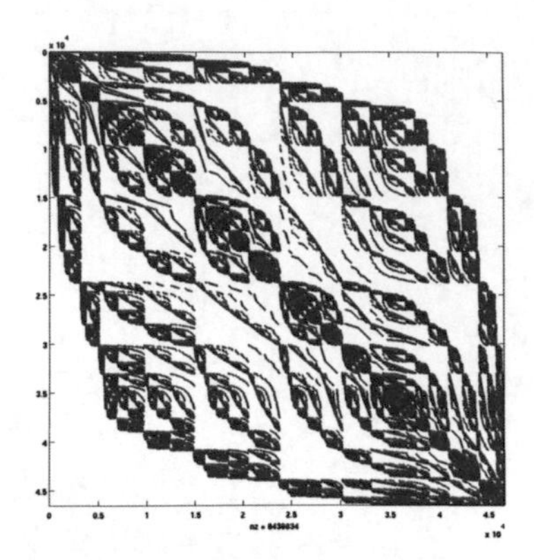

Figure D.2.: The matrix structure of the LC Schwinger model Hamiltonian at harmonic resolution $K = 40$. Blue dots indicate the 8438834 non-vanishing matrix elements. The matrix size is roughly $5 \cdot 10^4 \times 5 \cdot 10^4$.

one can imagine three arrays containing line, row indices, and value of the non-zero elements. The advantage of the sparse format is the reduced memory demand. The price to be paid is the less efficient implementation of matrix operations especially the multiplication of two matrices. We used the sparse matrix routines build in MATLAB which have been developed around the LAPACK program package. However, the matrix dimensions beyond $K = 55$ are too large to probe these regimes with present personal computers. For the computation of the lower eigenvalues we used the symmetric Lanzcos algorithm which approximates the eigenvalues and the corresponding eigenvectors of largest modulus. Our description of the Lanzcos algorithm follows the textbook [130]. Compared to transformation methods like QR, Jacobi or Householder transformations, the Lanzcos and the related power method are iterative algorithms. This class of algorithms can give partial informations about the eigenspectrum of large (sparse) matrices without constructing the unitary transformation to the eigenbasis explicitly. A Lanzcos diagonalization includes two steps in general. First a tridiagonal matrix is iteratively constructed having the same spectrum as the original matrix. Second, QR or other transformation methods are used to determine the eigenvalues of the tridiagonal matrix. Due to the tridiagonal form the QR algorithm terminates fast. The tridiagonalization step will be explained in more detail. Let A denote some arbitrary symmetric matrix. By the Cayley-Hamilton theorem the matrix A is solution to its own characteristic equation, i.e.

$$c_1 I + c_2 A + c_3 A^2 + \ldots c_n A^{n-1} + A^n = 0 \tag{D.6}$$

where the characteristic equation being given by (D.6) with A replaced by some generic eigenvalue λ. By choosing some arbitrary non-zero vector v_1 one constructs a Krylov sequence

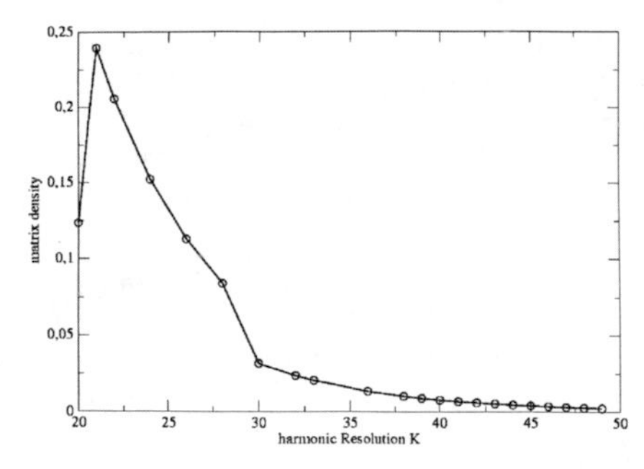

Figure D.3.: The matrix density, i.e. the ratio of non-vanishing elements to the total size of the matrix, as a function of the harmonic resolution.

$\{v_1, v_2, \ldots, v_n\}$ via

$$v_{i+1} = Av_i. \tag{D.7}$$

Multiplying (D.6) by v_1 from the right one finds

$$c_1 v_1 + c_2 v_2 + \ldots c_n v_n + Av_n = 0 \tag{D.8}$$

relating the vectors in the Krylov sequence and Av_n. One may merge (D.7) and (D.8) into the matrix equation

$$AV = VC \qquad \text{with} \qquad C = \begin{pmatrix} 0 & & & & -c_1 \\ 1 & 0 & & & -c_2 \\ & 1 & \ddots & & \vdots \\ & & \ddots & 0 & -c_{n-1} \\ & & & 1 & -c_n \end{pmatrix}, \tag{D.9}$$

where the Krylov vector v_i is the i-th row in the matrix V. In principle, one can compute V, then solve (D.9) for C and find the zeros of the characteristic polynomial. As seen in the power method the vectors v_i become more and more parallel and thus numerically unstable. In the standard form one supplements the algorithm by additionally orthogonalizing the vectors in the Krylov sequence. Let $Y = \{y_1, y_2, \ldots, y_n\}$ be a set of mutually orthogonal vectors spanning the same vector space as the v's. Then equation (D.9) is transformed to

$$AY = YT \qquad \text{with} \qquad Y^\dagger Y = I, \tag{D.10}$$

where T is symmetric and tridiagonal, i.e. of the form

$$T = \begin{pmatrix} \alpha_1 & \beta_1 & & \\ \beta_1 & \alpha_2 & \ddots & \\ & \ddots & \ddots & \beta_{n-1} \\ & & \beta_{n-1} & \alpha_n \end{pmatrix}. \tag{D.11}$$

In practical applications one can start with some arbitrary non-zero vector y_1 and force the mutual orthogonality of the y's during their construction. The mutual orthogonality can get lost due to numerical rounding errors, in particular, when the original matrix is bad-conditioned, i.e. the ratio of the smallest and the largest eigenvalue, the so-called condition number, $c = {|\lambda_1|}/{|\lambda_n|}$ is large. Besides the vectors y one determines all elements of T iteratively from (D.10).

For large matrices one is able to determine the eigenvalues of large modulus easily and with high precision applying the Lanzcos algorithms. We are however interested in the small eigenvalues of the LC mass matrix. Because the spectrum of the LC Hamiltonian is bounded from below we shift the matrix by the largest eigenvalue, that means, we use Lanzcos algorithm to compute the eigenvalues of $A - \lambda_n I$ instead of A. Afterwards the shift is undone by adding λ_n to the obtained eigenvalues.

D.3. Traces of matrix exponentials

There are various ways of computing the matrix exponential. In this work we want to obtain the trace of a matrix exponential as this quantity presents the partition function which is central for the study of thermodynamical properties. Roughly speaking, two possibilities exist.

Since we are only interested in the trace one can compute the matrix exponential in any basis and sum the diagonal elements. Here approximations like the truncation of the Taylor series for the matrix exponential have to be introduced. The second option is to compute the (or a subset of the) eigenvalues and take the sum of their exponentials until the further contribution is small. Compared to the other suggestion the matrix exponential computed via the eigenvalues is the most accurate method but is of course not feasible for large matrices. Therefore we stay with the former option. But instead of computing the exponentiated matrix explicitly we used a random vector routine [131, 132] as indicated in Section 4.4 to approximate its trace. The estimated trace reads

$$\langle \mathrm{Tr}\; e^{\beta P^0_{LC}} \rangle = \frac{1}{S} \sum_{p=1}^{S} \sum_{n,m=1}^{D} \bar{c}_{m,p} c_{n,p} \, \langle \phi_m | e^{\beta P^0_{LC}} | \phi_n \rangle, \tag{D.12}$$

where S, D denote the sample size and matrix dimension correspondingly. The coefficients c_n (and their complex conjugates $\bar{c}_n$) are in general complex random variables (i.e. two real ones) with some distribution and the vectors $|\phi_n\rangle$ fix some matrix representation of the exponential. In the DLCQ case the vectors are free particle Fock vectors. The statistical distribution of c_n and $\bar{c}_n$ is chosen such that one has the following relations for the moments

$$\langle\langle c_n \rangle\rangle = \langle\langle \bar{c}_n \rangle\rangle = 0, \tag{D.13}$$

$$\langle\langle \bar{c}_m c_n \rangle\rangle = \delta_{m,n}, \tag{D.14}$$

where $\langle\langle \cdot \rangle\rangle$ stands for statistical averaging. These properties ensure the convergence by observing

$$\langle\Phi|A|\Phi\rangle = \sum_n A_{nn} + (\bar{c}_n c_m - \delta_{m,n})\, A_{nm}. \tag{D.15}$$

By averaging (D.15) the second term vanishes and thus establishes the convergence with the rate of order $1/\sqrt{S}$. If one constrains the random vectors to the unit sphere even faster convergences of order $O(1/S)$ is achieved [131]. Yet there is the possibility to fix what distribution the random vectors are drawn from. This distribution shall be symmetric under exchange of two components and an even function in each component. Most easily one assumes that each component of the random vector has the same even statistical distribution. The distribution of the vector is then simply the product of the single distributions. The simplest even distribution is the uniform one.

In practical computations the special case of real random sign vectors with (automatically) uniform distributions was used to compute the traces (D.12). In the case of real random vectors we draw each (real) component out of the set $\{-1,1\}$ with equal probability and then normalize the vector. These random vectors are exact for diagonal matrices, meaning that statistical fluctuations are absent. In the computations with the DLCQ Hamiltonians we found that averaging 100 random vectors was sufficient to obtain accurate results in the considered range of harmonic resolutions. A detailed analysis of the statistical errors is missing so far but for the results presented in Section 4.4 we expect other error sources, e.g. the extrapolation error, to outweigh the statistical error introduced here.

Because the LC Hamiltonian is nearly diagonal for large coupling $m/g \gg 1$ we used the Trotter decomposition to split H_{LC} in P^0_{LC} into the free Hamiltonian and the off-diagonal interaction. Trotter approximants for a general Hamiltonian $H = H_0 + V$ are usually encountered by writing the partition function as

$$\mathcal{Z} = \mathrm{Tr}\left[\left(\exp\{-\frac{\beta}{k}(H_0+V)\}\right)^k\right], k \in \mathbb{N}. \tag{D.16}$$

The lowest order (hermitian) Trotter approximation for the exponential operator is

$$e^{-\tau(H_0+V)} = e^{-\frac{\tau}{2}H_0} e^{-\tau V} e^{-\frac{\tau}{2}H_0} + O(\tau^3), \tag{D.17}$$

where $\tau = \beta L/2\pi k$. Inserting the expansion (D.17) in Eq. (D.16) the full partition function is recovered in the limit $k \longrightarrow \infty$. Higher order approximants can be found, e.g. in Ref. [133], and one has to fifth order in τ

$$e^{-\tau(H_0+V)} = e^{-\frac{\tau}{2}H_0} e^{-\frac{\tau}{2}V} e^{-\frac{\tau^3}{4}C} e^{-\frac{\tau}{2}V} e^{-\frac{\tau}{2}H_0} + O(\tau^5), \tag{D.18}$$

with $C = [[V, H_0], H_0 + 2V]/2$. In our numerical implementation we only used the lowest order Trotter approximation so far. Another factor of 2 is gained by utilizing that H is hermitian as

$$\mathcal{Z} = \sum_n \langle\phi_n|e^{-\beta H}|\phi_n\rangle = \sum_n \langle e^{-\frac{\beta}{2}H}\phi_n|e^{-\frac{\beta}{2}H}\phi_n\rangle. \tag{D.19}$$

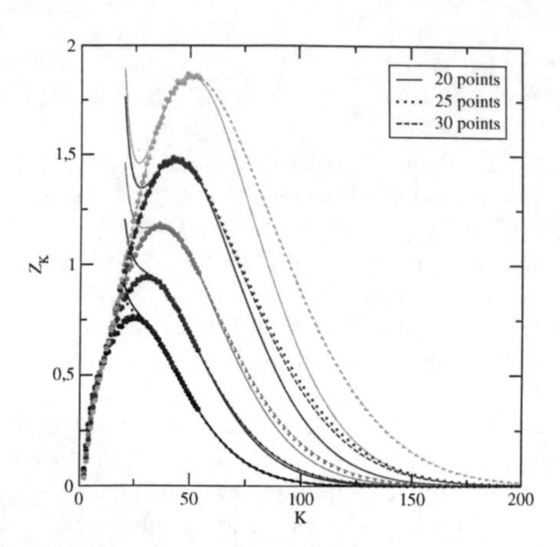

Figure D.4.: The function $\mathcal{Z}_K$ for the coupling $m/g = 1$ and the temperature $T = 0.5M_0$. The various colors, black (red, green, blue, orange), stand for different volume sizes $L/(2\pi M_0) = 550$ $(600, 650, 700, 750)$, correspondingly. The number of points used in the fit routine are indicated in the legend. The numerical results are depicted by dots.

One needs to propagate the random vector only by $\beta/2$ in imaginary time and calculate its norm at the end.

Combining the two methods above, and leaving the trivial exponential of P^+ aside, the norm of the vector

$$e^{-\frac{\beta}{2}\frac{L}{2\pi}(H_0+V)}|\phi\rangle = e^{-\frac{\beta}{4}\frac{L}{2\pi}H_0}\left(\sum_{j=1}^{J}\frac{(-1)^j}{j!}(\frac{\beta}{2}\frac{L}{2\pi}V)^j\right)e^{-\frac{\beta}{4}\frac{L}{2\pi}H_0}|\phi\rangle, \tag{D.20}$$

was computed S times where $|\phi\rangle$ is a generic random vector. The first and third exponents are easily computed exactly. For the interaction term the parameter J was tuned such that additional terms in the Taylor expansion are negligible. One advantage of (D.20) is that only matrix-vector operations are necessary, so multiplying two matrices, which is computational costly in the sparse format, is omitted completely.

Although the presented methods are very powerful, one is limited by the maximal resolution available in the DLCQ application. For increasing temperatures or volumes the upper bound of $K = 55$ is exceeded and some kind of extrapolation is necessary. The lower part of the spectrum is converged at these high values of harmonic resolution and one may expect to just take these mass eigenvalues and compute the exponential (4.92) to arbitrary K. Proceeding that way leads to a large underestimation of the partition function because the exponentially growing number of states is neglected. Since it is known how the number of partitions depends on K, Eq. (D.5), an educated guess for an extrapolation function incorporates this dependence. For very high K the suppression factor $\exp\{-\beta P^+\}$ overtakes and $\mathcal{Z}_K$, the contribution of the K Fock sector to the partition function, tends toward zero. Thus, we tried a five-parameter fit to the last 10 to 30 data points of $\mathcal{Z}_K$ and found that an extrapolation to high K values works surprisingly well. The function subject to a maximum likelihood fit is

$$f(K) = \exp\left\{\alpha_0 + \alpha_{-1}/K + \alpha_{-2}/K^2 + \alpha_1 K + \alpha_2 K^2\right\} \tag{D.21}$$

Using the MINUIT routine one introduces small artificial error bars to the data points and determines the coefficients $\alpha_i,\ i = -2, \ldots, 2$. A typical situation is depicted in Figure D.4. The extrapolation has to be done for every tupel of external parameters (T, L). In Figure D.4 the temperature is $T/M_0 = 0.5$ and the plotted volumes range from $L/(2\pi M_0) = 550$ (black curve) to $L/(2\pi M_0) = 750$ (orange curve). The aim of the fit is the $\mathcal{Z}_K$ distribution for large resolutions, i.e. to the right of the maximum. If sufficiently many data points are located beyond the maximum the fit routine give stable results under varying the number of fit points, see e.g. the black and red curves in Figure D.4. For increasing volumes the fit becomes unstable because the data points to be fitted are mostly located left of the maximum. This is the case for the blue and orange curves in Fig. D.4 but the differences between the fit functions are suppressed in the thermodynamical potential by taking the logarithm. The start values of the parameters α_i are important for the performance of the fit procedure. Instead of taking fixed start values we changed these inputs continuously following the variation of the external parameters T, L.

E. Linear scaling of $\ln \mathcal{Z}$ - interacting system

Here we show an intermediate result in the determination of the thermodynamical potential in Section 4.4. This is analog to the Figure 3.2 for the ideal gas. One has to compute $\ln \mathcal{Z}$ for each temperature and several box sizes. Any of these values should be converged in K, then the bulk limit $L \to \infty$ can be addressed. In Figure E.1 we have depicted the scaling behavior of $\ln \mathcal{Z}$ in L. The panels in Fig. E.1 range from $T/M_0 = 0.3$ in (A) to $T/M_0 = 1.2$ in (J) increasing in steps of $\Delta T = 0.1 M_0$. The black curve is the original data as resulting from the DLCQ approximation (4.92). The blue and brown curves stem from the extrapolation taking 10 or 30 points into account.

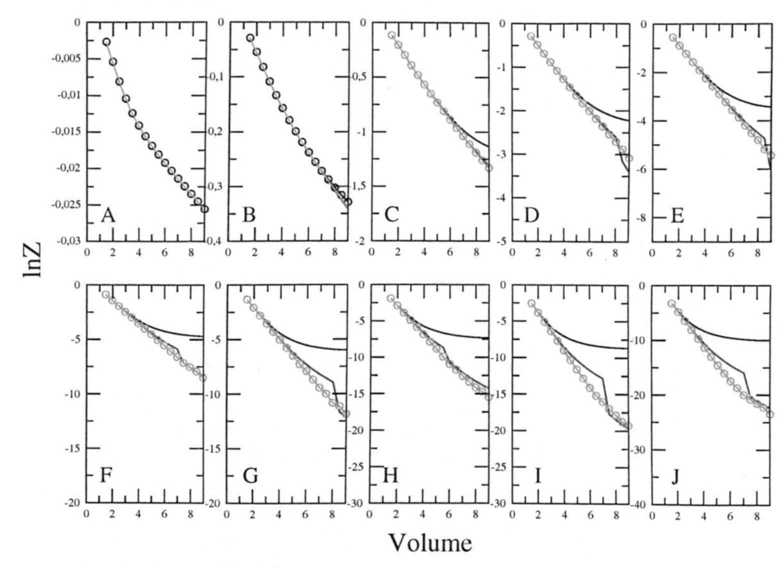

Figure E.1.: The thermodynamical potential as a function of the volume for different temperatures. See the text for further explanations.

Bibliography

[1] C. R. Allton et al. QCD at non-zero temperature and density from the lattice. *Nucl. Phys. Proc. Suppl.*, 141:186–190, 2005.

[2] Hunting the Quark Gluon Plasma: Results from the first 3 Years at RHIC. *Nuclear Physics A*, 757(1-2):1–283, 2005.

[3] P. Huovinen and P. V. Ruuskanen. Hydrodynamic Models for Heavy Ion Collisions. *Ann. Rev. Nucl. Part. Sci.*, 56:163–206, 2006.

[4] Matthias Troyer and Uwe-Jens Wiese. Computational complexity and fundamental limitations to fermionic quantum Monte Carlo simulations. *Phys. Rev. Lett.*, 94:170201, 2005.

[5] Paul A. M. Dirac. Forms of relativistic dynamics. *Rev. Mod. Phys.*, 21:392–399, 1949.

[6] ILCAC Whitepaper. Light-Front Qantum Chromodynamics - A framework for analysis of hadron dynamics. to be found at http://www.ilcacinc.org/.

[7] Kenneth G. Wilson. The origins of lattice gauge theory. *Nucl. Phys. Proc. Suppl.*, 140:3–19, 2005.

[8] H. Arthur Weldon. Thermal field theory and generalized light front quantization. *Phys. Rev.*, D67:085027, 2003.

[9] Toshihide Maskawa and Koichi Yamawaki. The problem of p+ = 0mode in the null plane field theory and dirac's method of quantization. *Prog. Theor. Phys.*, 56:270, 1976.

[10] Stanley J. Brodsky, Hans-Christian Pauli, and Stephen S. Pinsky. Quantum chromodynamics and other field theories on the light cone. *Phys. Rept.*, 301:299–486, 1998.

[11] Matthias Burkardt. Light front quantization. *Adv. Nucl. Phys.*, 23:1–74, 1996.

[12] Matthias Burkardt and Simon Dalley. The relativistic bound state problem in qcd: Transverse lattice methods. *Prog. Part. Nucl. Phys.*, 48:317–362, 2002.

[13] Thomas Heinzl. Light cone dynamics of particles and fields. *Habilitation thesis*, 1998.

[14] H. Leutwyler and J. Stern. Relativistic dynamics on a null plane. *Ann. Phys.*, 112:94, 1978.

[15] D. Gromes, H. J. Rothe, and B. Stech. Field quantization on the surface x-squared = constant. *Nucl. Phys.*, B75:313–332, 1974.

[16] K. Bardakci and M. B. Halpern. Theories at infinite momentum. *Phys. Rev.*, 176:1686–1699, 1968.

[17] Steven Weinberg. Dynamics at infinite momentum. *Phys. Rev.*, 150:1313–1318, 1966.

[18] E Minguzzi. Classical aspects of lightlike dimensional reduction. *Classical and Quantum Gravity*, 23(23):7085–7110, 2006.

[19] F. Lenz, M. Thies, K. Yazaki, and S. Levit. Hamiltonian formulation of two-dimensional gauge theories on the light cone. *Annals Phys.*, 208:1–89, 1991.

[20] T. Heinzl, H. Kroger, and N. Scheu. Loss of causality in discretized light-cone quantization. 1999.

[21] A. Harindranath, L. Martinovic, and J. P. Vary. Compactification near and on the light front. *Phys. Rev.*, D62:105015, 2000.

[22] D. Grunewald, E. M. Ilgenfritz, E. V. Prokhvatilov, and H. J. Pirner. Formulating Light Cone QCD on the Lattice. *Phys. Rev.*, D77:014512, 2008.

[23] L. D. Faddeev and R. Jackiw. Hamiltonian Reduction of Unconstrained and Constrained Systems. *Phys. Rev. Lett.*, 60:1692, 1988.

[24] J. W. Jun and C. K. Jue. Faddeev-Jackiw quantization and the light cone zero mode problem. *Phys. Rev.*, D50:2939–2941, 1994.

[25] V. Parameswaran Nair. *Quantum Field Theory: A Modern Perspective.* Springer, 2005.

[26] T. Heinzl, C. Stern, E. Werner, and B. Zellermann. The vacuum structure of light-front phi**4(1+1)-theory. *Z. Phys.*, C72:353–364, 1996.

[27] George Leibbrandt. Introduction to Noncovariant Gauges. *Rev. Mod. Phys.*, 59:1067, 1987.

[28] Hans Christian Pauli and Stanley J. Brodsky. Solving field theory in one space one time dimension. *Phys. Rev.*, D32:1993, 1985.

[29] Hans Christian Pauli and Stanley J. Brodsky. Discretized light cone quantization: Solution to a field theory in one space one time dimensions. *Phys. Rev.*, D32:2001, 1985.

[30] Thomas Eller, Hans Christian Pauli, and Stanley J. Brodsky. Discretized light cone quantization: The massless and the massive schwinger model. *Phys. Rev.*, D35:1493, 1987.

[31] Kent Hornbostel, Stanley J. Brodsky, and Hans Christian Pauli. Light cone quantized qcd in (1+1)-dimensions. *Phys. Rev.*, D41:3814, 1990.

[32] O. Lunin and S. Pinsky. SDLCQ: Supersymmetric discrete light cone quantization. *AIP Conf. Proc.*, 494:140–218, 1999.

[33] Sho Tsujimaru and Koichi Yamawaki. Zero mode and symmetry breaking on the light front. *Phys. Rev.*, D57:4942–4964, 1998.

[34] Lubomir Martinovic and Marshall Luban. Analytic solution of the microcausality problem in discretized light cone quantization. *Phys. Lett.*, B605:203–213, 2005.

[35] Dipankar Chakrabarti, Asmita Mukherjee, Rajen Kundu, and A. Harindranath. A numerical experiment in DLQC: Microcausality, continuum limit and all that. *Phys. Lett.*, B480:409–417, 2000.

[36] Andrew C. Tang, Stanley J. Brodsky, and Hans Christian Pauli. Discretized light cone quantization: Formalism for quantum electrodynamics. *Phys. Rev.*, D44:1842–1865, 1991.

[37] M. Krautgartner, H. C. Pauli, and F. Wolz. Positronium and heavy quarkonia as testing case for discretized light cone quantization. 1. *Phys. Rev.*, D45:3755–3774, 1992.

[38] William A. Bardeen, Robert B. Pearson, and Eliezer Rabinovici. Hadron Masses in Quantum Chromodynamics on the Transverse Lattice. *Phys. Rev.*, D21:1037, 1980.

[39] Paul A. Griffin. Solving (3+1) QCD on the transverse lattice using (1+1) conformal field theory. *Nucl. Phys.*, B372:270–292, 1992.

[40] M. Burkardt. Light front ensemble projector monte carlo. *Phys. Rev.*, D49:5446–5457, 1994.

[41] S. Dalley and B. van de Sande. Transverse lattice approach to light front Hamiltonian QCD. *Phys. Rev.*, D59:065008, 1999.

[42] Carl M. Bender, Stephen Pinsky, and Brett Van de Sande. Spontaneous symmetry breaking of Phi**4 in (1+1)- dimensions in light front field theory. *Phys. Rev.*, D48:816–821, 1993.

[43] Stephen S. Pinsky and Brett van de Sande. Spontaneous symmetry breaking of (1+1)-dimensional phi**4 theory in light front field theory. 2. *Phys. Rev.*, D49:2001–2013, 1994.

[44] Stephen S. Pinsky, Brett van de Sande, and John R. Hiller. Spontaneous symmetry breaking of (1+1)-dimensional phi**4 theory in light front field theory. 3. *Phys. Rev.*, D51:726–733, 1995.

[45] T. Heinzl, S. Krusche, and E. Werner. Spontaneous symmetry breaking in light cone quantum field theory. *Phys. Lett.*, B272:54–60, 1991.

[46] T. Heinzl, S. Krusche, S. Simburger, and E. Werner. Nonperturbative light cone quantum field theory beyond the tree level. *Z. Phys.*, C56:415–420, 1992.

[47] Takanori Sugihara and Masa-aki Taniguchi. Manifestation of a nontrivial vacuum in discrete light cone quantization. *Phys. Rev. Lett.*, 87:271601, 2001.

[48] S. S. Chabysheva and J. R. Hiller. Zero momentum modes in discrete light-cone quantization. 2009.

[49] Aiichi Iwazaki. Spontaneous breakdown of U(1) symmetry in DLCQ without zero mode. *Phys. Rev.*, D75:105009, 2007.

[50] Joel S. Rozowsky and Charles B. Thorn. Spontaneous symmetry breaking at infinite momentum without P+ zero modes. *Phys. Rev. Lett.*, 85:1614–1617, 2000.

[51] Dipankar Chakrabarti, A. Harindranath, Lubomir Martinovic, and J. P. Vary. Kinks in discrete light cone quantization. *Phys. Lett.*, B582:196–202, 2004.

[52] Dipankar Chakrabarti, A. Harindranath, Lubomir Martinovic, Grigorii B. Pivovarov, and James P. Vary. Ab initio results for the broken phase of scalar light front field theory. *Phys. Lett.*, B617:92–98, 2005.

[53] Dipankar Chakrabarti, A. Harindranath, and J. P. Vary. A transition in the spectrum of the topological sector of phi**4(2) theory at strong coupling. *Phys. Rev.*, D71:125012, 2005.

[54] David E. Miller and Frithjof Karsch. Covariant Structure of Relativistic Gases in Equilibrium. *Phys. Rev.*, D24:2564, 1981.

[55] H. Arthur Weldon. Covariant Calculations at Finite Temperature: The Relativistic Plasma. *Phys. Rev.*, D26:1394, 1982.

[56] Jorg Raufeisen and Stanley J. Brodsky. Statistical physics and light-front quantization. *Phys. Rev.*, D70:085017, 2004.

[57] A N. Kvinikhidze and B. Blankleider. Equivalence of light-front and conventional thermal field theory. *Phys. Rev.*, D69:125005, 2004.

[58] H. Arthur Weldon. Thermal self-energies using light-front quantization. *Phys. Rev.*, D67:128701, 2003.

[59] Ashok K. Das and Silvana Perez. Quantization in a general light-front frame. *Phys. Rev.*, D70:065006, 2004.

[60] Ashok Das and Xing-xiang Zhou. Light-front schwinger model at finite temperature. *Phys. Rev.*, D68:065017, 2003.

[61] S. Dalley and B. van de Sande. Finite temperature gauge theory from the transverse lattice. *Phys. Rev. Lett.*, 95:162001, 2005.

[62] John R. Hiller, Yiannis Proestos, Stephen Pinsky, and Nathan Salwen. N = (1,1) super yang-mills theory in 1+1 dimensions at finite temperature. *Phys. Rev.*, D70:065012, 2004.

[63] M. Beyer, S. Mattiello, T. Frederico, and H. J. Weber. Light-front field theory of quark matter at finite temperature in the Nambu-Jona-Lasinio model. *J. Phys.*, G31:21–27, 2005.

[64] S. Strauss, S. Mattiello, and M. Beyer. Light-front Nambu–Jona-Lasinio model at finite temperature and density. *J. Phys.*, G36:085006, 2009.

[65] S. Strauss and M. Beyer. Light front QED_{1+1} at finite temperature. *Phys. Rev. Lett.*, 101:100402, 2008.

[66] John R. Hiller, Stephen Pinsky, Yiannis Proestos, Nathan Salwen, and Uwe Trittmann. Spectrum and thermodynamic properties of two-dimensional N = (1,1) super Yang-Mills theory with fundamental matter and a Chern-Simons term. *Phys. Rev.*, D76:045008, 2007.

[67] E. Landau, L.and Lifshitz. *Statistical Physics part 1*. Pergamon Press, 1980.

[68] Gernot Neugebauer. *Relativistische Thermodynamik (in german)*. Akademie-Verlag Berlin, 1980.

[69] A. Das. *Finite Temperature Field Theory*. World Scientific, 1999.

[70] V. S. Alves, Ashok Das, and Silvana Perez. Light-front field theories at finite temperature. *Phys. Rev.*, D66:125008, 2002.

[71] M. Beyer, S. Mattiello, T. Frederico, and H. J. Weber. Three-quark clusters at finite temperatures and densities. *Phys. Lett.*, B521:33–41, 2001.

[72] S. Lenz. Statistische Mechanik auf dem Lichtkegel. Diploma thesis, unpublished, in german, Erlangen-Nuernberg University, 1990.

[73] B. Blankleider and A. N. Kvinikhidze. Comment on 'Light-front Schwinger model at finite temperature'. *Phys. Rev.*, D69:128701, 2004.

[74] Ashok K. Das and Xing-xiang Zhou. Reply to 'Comment on 'Light-front Schwinger model at finite temperature' '. *Phys. Rev.*, D69:128702, 2004.

[75] Sidney Coleman, R. Jackiw, and Leonard Susskind. Charge shielding and quark confinement in the massive schwinger model. *Annals of Physics*, 93(1-2):267 – 275, 1975.

[76] Sidney R. Coleman. More About the Massive Schwinger Model. *Ann. Phys.*, 101:239, 1976.

[77] P. Sriganesh, R. Bursill, and C. J. Hamer. A new finite-lattice study of the massive schwinger model. *Phys. Rev.*, D62:034508, 2000.

[78] Julian S. Schwinger. Gauge Invariance and Mass. 2. *Phys. Rev.*, 128:2425–2429, 1962.

[79] Stephan Elser and Alex C. Kalloniatis. Qed(1+1) at finite temperature – a study with light-cone quantisation. *Phys. Lett.*, B375:285–291, 1996.

[80] T. Byrnes, P. Sriganesh, R. J. Bursill, and C. J. Hamer. Density matrix renormalisation group approach to the massive schwinger model. *Phys. Rev.*, D66:013002, 2002.

[81] J. H. Lowenstein and J. A. Swieca. Quantum electrodynamics in two-dimensions. *Ann. Phys.*, 68:172–195, 1971.

[82] A. Casher, John B. Kogut, and Leonard Susskind. Vacuum polarization and the absence of free quarks. *Phys. Rev.*, D10:732–745, 1974.

[83] Sidney R. Coleman. Quantum sine-Gordon equation as the massive Thirring model. *Phys. Rev.*, D11:2088, 1975.

[84] Max A. Metlitski. Is Schwinger model at finite density a crystal? *Phys. Rev.*, D75:045004, 2007.

[85] Kerson Huang. *Quarks, Leptons and Gauge Fields*. Singapore, Singapore: World Scientific, 1982.

[86] G. Mccartor. Light Cone Quantuzation for massless Fields. *Z. Phys.*, C41:271–275, 1988.

[87] Lubomir Martinovic. Large gauge transformations and the light-front vacuum structure. *Phys. Lett.*, B509:355–364, 2001.

[88] A. C. Kalloniatis and H. C. Pauli. Bosonic zero modes and gauge theory in discrete light cone quantization. *Z. Phys.*, C60:255–264, 1993.

[89] L. Martinovic. Non-trivial Fock vacuum of the light front Schwinger model. *Phys. Lett.*, B400:335–340, 1997.

[90] L'ubomir Martinovic and James P. Vary. Theta-vacuum of the bosonized massive light-front Schwinger model. *Phys. Lett.*, B459:186–192, 1999.

[91] Alex C. Kalloniatis and David G. Robertson. Theta Vacua in the Light-Cone Schwinger Model. *Phys. Lett.*, B381:209–215, 1996.

[92] T. Heinzl, S. Krusche, and E. Werner. The Fermionic Schwinger model in light cone quantization. *Phys. Lett.*, B275:410–418, 1992.

[93] Koji Harada, Atsushi Okazaki, and Masa-aki Taniguchi. Dynamics of the Light-Cone Zero Modes: Theta Vacuum of the Massive Schwinger Model. *Phys. Rev.*, D55:4910–4919, 1997.

[94] T. Heinzl, S. Krusche, and E. Werner. Nontrivial vacuum structure in light cone quantum field theory. *Phys. Lett.*, B256:55–59, 1991.

[95] T. Heinzl. Fermion condensates and the trivial vacuum of light-cone quantum field theory. *Phys. Lett.*, B388:129–136, 1996.

[96] C. M. Yung and C. J. Hamer. Discretized light cone quantization of (1+1)-dimensional qed reexamined. *Phys. Rev.*, D44:2598–2601, 1991.

[97] Yi-zhang Mo and Robert J. Perry. Basis function calculations for the massive schwinger model in the light front tamm-dancoff approximation. *J. Comput. Phys.*, 108:159–174, 1993.

[98] Helmut Kroger and Norbert Scheu. The massive schwinger model: A hamiltonian lattice study in a fast moving frame. *Phys. Lett.*, B429:58–63, 1998.

[99] Brett van de Sande. Convergence of discretized light cone quantization in the small mass limit. *Phys. Rev.*, D54:6347–6350, 1996.

[100] Koji Harada, Thomas Heinzl, and Christian Stern. Variational mass perturbation theory for light-front bound-state equations. *Phys. Rev.*, D57:2460–2474, 1998.

[101] Kent Hornbostel. *The application of light cone quantization to quantum chromodynamics in (1+1)-dimensions*. PhD thesis, Stanford - SLAC, 1988.

[102] C. Moler and C. Van Loan. Nineteen Dubious Ways to Compute the Exponential of a Matrix, Twenty-Five Years Later. *SIAM Review*, 45:3–000, 2003.

[103] P. Gonzalez and V. Vento. Color Singlet States in a Hadronic Quarl Cluster Basis. *Acta Phys. Austriaca*, 2:145–154, 1987.

[104] Stanisław D. Głazek and Kenneth G. Wilson. Renormalization of hamiltonians. *Phys. Rev. D*, 48(12):5863–5872, Dec 1993.

[105] Franz Wegner. Flow-equations for Hamiltonians. *Annalen der Physik*, 506:77–91, 1994.

[106] F. Wegner. Flow equations and normal ordering: a survey. *Journal of Physics A Mathematical General*, 39:8221–8230, June 2006.

[107] Kenneth G. Wilson. The renormalization group: Critical phenomena and the kondo problem. *Rev. Mod. Phys.*, 47(4):773–840, Oct 1975.

[108] Robert J. Perry. A Renormalization group approach to Hamiltonian light front field theory. *Ann. Phys.*, 232:116–222, 1994.

[109] Brent H. Allen and Robert J. Perry. Systematic renormalization in Hamiltonian light-front field theory. *Phys. Rev.*, D58:125017, 1998.

[110] Martina M. Brisudova, Robert J. Perry, and Kenneth G. Wilson. Quarkonia in Hamiltonian light-front QCD. *Phys. Rev. Lett.*, 78:1227–1230, 1997.

[111] Brent H. Allen and Robert J. Perry. Glueballs in a Hamiltonian light front approach to pure glue QCD. *Phys. Rev.*, D62:025005, 2000.

[112] E. L. Gubankova and F. Wegner. Flow equations for QED in the light front dynamics. *Phys. Rev.*, D58:025012, 1998.

[113] Steven R. White. Density matrix formulation for quantum renormalization groups. *Phys. Rev. Lett.*, 69(19):2863–2866, Nov 1992.

[114] Steven R. White. Strongly correlated electron systems and the density matrix renormalization group. *Physics Reports*, 301(1-3):187 – 204, 1998.

[115] M. A. Martin-Delgado and G. Sierra. A density matrix renormalization group approach to an asymptotically free model with bound states. *Phys. Rev. Lett.*, 83:1514–1517, 1999.

[116] U. Schollwöck. The density-matrix renormalization group. *Rev. Mod. Phys.*, 77(1):259–315, Apr 2005.

[117] Steven R. White. Density matrix renormalization group algorithms with a single center site. *Phys. Rev. B*, 72(18):180403, Nov 2005.

[118] F. Verstraete, V. Murg, and J. I. Cirac. Matrix product states, projected entangled pair states, and variational renormalization group methods for quantum spin systems. *Advances in Physics*, 57:143–224, 2008.

[119] Takanori Sugihara. Density matrix renormalization group in a two-dimensional lambda phi**4 Hamiltonian lattice model. *JHEP*, 05:007, 2004.

[120] T. Xiang. Density-matrix renormalization-group method in momentum space. *Phys. Rev. B*, 53(16):10445–10448, Apr 1996.

[121] Satoshi Nishimoto, Eric Jeckelmann, Florian Gebhard, and Reinhard M. Noack. Application of the density matrix renormalization group in momentum space. *Phys. Rev. B*, 65(16):165114, Apr 2002.

[122] Xiaoqun Wang and Tao Xiang. Transfer-matrix density-matrix renormalization-group theory for thermodynamics of one-dimensional quantum systems. *Phys. Rev. B*, 56(9):5061–5064, Sep 1997.

[123] Yutaka Hosotani. Massive multi-flavor Schwinger model at finite temperature and on compact space. 1995.

[124] Matthias Burkardt and Koji Harada. Theta dependence of meson masses in the small mass limit of the massive Schwinger model. *Phys. Rev.*, D57:5950–5954, 1998.

[125] J. P. Vary et al. Hamiltonian light-front field theory in a basis function approach. 2009.

[126] Dimitra Karabali and V. P. Nair. A gauge-invariant Hamiltonian analysis for non-Abelian gauge theories in (2+1) dimensions. *Nucl. Phys.*, B464:135–152, 1996.

[127] Georg M. von Hippel and Mattias N. R. Wohlfarth. Covariant canonical quantization. *Eur. Phys. J.*, C47:861–872, 2006.

[128] M. Heyssler and A. C. Kalloniatis. Constituent quark picture out of QCD in two-dimensions on the light cone. *Phys. Lett.*, B354:453–459, 1995.

[129] George E. Andrews. *The theory of partitions*. Cambridge University Press, 1998.

[130] Alan Jennings and J.J. McKeown. *Matrix Computation - 2nd ed.* John Wiley & Sons, 1992.

[131] Anthony Hams and Hans De Raedt. Fast algorithm for finding the eigenvalue distribution of very large matrices. *Phys. Rev. E*, 62(3):4365–4377, Sep 2000.

[132] Toshiaki Iitaka and Toshikazu Ebisuzaki. Random phase vector for calculating the trace of a large matrix. *Phys. Rev. E*, 69(5):057701, May 2004.

[133] Hans De Raedt and Bart De Raedt. Applications of the generalized trotter formula. *Phys. Rev. A*, 28(6):3575–3580, Dec 1983.

Bibliography

Acknowledgments

I am indebted to PD Dr. Michael Beyer for suggesting the interesting research topic as well as for his sustained encouragement during the years. Furthermore I thank Professor Stanley Brodsky and Professor Tobias Frederico for serving as the referees of my thesis.

I have benefited from the interaction with many members of the light cone research community during the yearly meetings and appreciate their continuous interests in this work.

The inspiring atmosphere of the Elementary Particle Physics Group of Professor Henning Schröder is gratefully acknowledged. In particular, I would like to express my gratitude to Erik Schmidt, Dr. Stefano Mattiello and Robert Steinbeiß for discussing physics and for proof-reading of the script.

Throughout the project the untiring support of my family – specifically my mother Marianne, my father Raimond and my sister Christine – were invaluable. Finally, I wish to thank my girlfriend Elin Jannermann for her understanding of the time and focus I required to complete this thesis.

This work was supported by the Deutsche Forschungsgemeinschaft under contract number BE1092/13.

Erklärung

Ich versichere hiermit an Eides statt, dass ich die vorliegende Arbeit selbständig angefertigt und ohne fremde Hilfe verfasst habe, keine außer den von mir angegebenen Hilfsmitteln und Quellen dazu verwendet habe und die den benutzten Werken inhaltlich und wörtlich entnommenen Stellen als solche kenntlich gemacht habe.

Rostock, January 20, 2010

Stefan Strauß

CURRICULUM VITAE

Name	**Stefan Strauss**
Address	Richard-Wagner-Strasse 18, D-18055 Rostock, Germany (private)
	Institut für Physik, Universitätsplatz 3, D-18051 Rostock Phone: +49 381 4986776, Fax: +49 381 4986772 email: stefan.strauss@uni-rostock.de
date of birth	04/04/1980
place of birth	Ludwigslust, Germany

EDUCATION AND DEGREES

10/2009	Ph.D. degree "summa cum laude" on "Thermodynamics of low-dimensional light front gauge theories"
2005 - 2009	Ph.D. student in physics at University of Rostock funded by the German Research Association (Deutsche Forschungsgemeinschaft) under supervision of PD Dr. M. Beyer
08/2005	Physics master degree (Diplom Physiker), "excellent" (best possible grade), title of thesis "Two-body correlations in the light-front NJL model at finite temperature and density" (written in German)
1999 - 2005	Studies of physics at University of Rostock
1998 - 1999	Civilian service at Moorbad Bad Doberan
07/1998	Abitur ("1.5", with "1" being the best possible grade), qualification for university entrance

RESEARCH STAYS

10/03 - 11/03	Internship at DESY, Hamburg under supervision of Prof. Wolfgang Kilian
03/06 - 05/06	Participation in the ECT* Doctoral training program "Computational Techniques in Strongly Interacting Systems" in Trento, Italy
08/06 - 10/06	Research stay with Prof. Tobias Frederico at Centro Técnico Aeroespacial, Instituto Tecnológico de Aeronáutica, Dep. de Física, São José dos Campos, São Paulo, Brazil

List of Pubblications

peer reviewed Journal Articles

1. M. Beyer, S. Strauss, P. Schuck and S. A. Sofianos,
   "Light clusters in nuclear matter of finite temperature",
   Eur. Phys. J. A **22** (2004) 261 [arXiv:nucl-th/0310055].

2. S. Strauss and M. Beyer,
   "Light front QED_{1+1} at finite temperature",
   Phys. Rev. Lett. **101**, 100402 (2008) [arXiv:0805.3147 [hep-th]].

3. S. Strauss and M. Beyer, *"Thermodynamical properties of QED in 1+1 dimensions within light front dynamics"*,
   PoS **LC2008** (2008) 010 [arXiv:0810.5385 [nucl-th]].

4. S. Strauss, S. Mattiello and M. Beyer, *"Light-front Nambu–Jona-Lasinio model at finite temperature and density"*,
   J. Phys. G **36** (2009) 085006 [arXiv:0903.5209 [hep-ph]].

Conference Proceedings

1. S. Strauß, M. Beyer and S. Mattiello,
   "Restoration of chiral symmetry in light-front finite temperature field theory",
   Few Body Syst. **36**, 231 (2005), [arXiv:nucl-th/0410007]

2. S. Mattiello, M. Beyer, S. Strauss, T. Frederico and H. J. Weber,
   "Properties of three-body states in hot and dense quark matter",
   AIP Conf. Proc. **768**, 269 (2005)

3. S. Strauss, M. Beyer, S. Mattiello, T. Frederico and H. J. Weber,
   "Two-body correlations in hot quark matter",
   AIP Conf. Proc. **768**, 272 (2005)

4. M. Beyer, S. Mattiello, S. Strauss, T. Frederico, H. J. Weber, P. Schuck and S. A. Sofianos,
   "Dynamics of few-body states in a medium",
   AIP Conf. Proc. **768**, 392 (2005)

5. M. Beyer, S. Mattiello, S. Strauss, T. Frederico and H. J. Weber,
   "Light front approach to hot and dense quark matter",
   AIP Conf. Proc. **775**, 139 (2005)

6. S. Strauss, M. Beyer and S. Mattiello,
   "Light front NJL model at finite temperature",
   AIP Conf. Proc. **775**, 212 (2005)

7. S. Strauss, M. Beyer and S. Mattiello,
   "Light front approach to correlations in hot quark matter",
   Acta Phys. Hung. A **27** (2006) 379, [arXiv:nucl-th/0511005].

8. M. Beyer, S. Mattiello and S. Strauss,
   "Light front field theory of relativistic quark matter",
   Acta Phys. Hung. A **27** (2006) 327, [arXiv:nucl-th/0511002].

9. S.Strauss, M. Beyer, S. Mattiello and T. Frederico
   "Relativistic quark matter in light front field theory",
   Nuclear Physics A, Volume 790, Issues 1-4, 627c-630c

10. S.Strauss and M. Beyer,
    "Massive Light front Schwinger Model at finite temperature",
    Prog. Part. Nucl. Phys. **62** (2009) 535.